Introduction to
WIND TURBINE ENGINEERING

Introduction to
WIND TURBINE ENGINEERING

by

Andrze J. Wortman

BUTTERWORTH PUBLISHERS
Boston • London
Sydney • Wellington • Durban • Toronto
An Ann Arbor Science Book

Ann Arbor Science is an imprint of Butterworth Publishers

Library of Congress Catalog Card Number 83-071593
ISBN0-250-40562-8

10 9 8 7 6 5 4 3 2 1

Published by Butterworth Publishers
10 Tower Office Park
Woburn, MA 01801

Printed in the United States of America

DEDICATION

To my parents, Stefan and Genevieve Wortman, to whose faith, support and encouragement I owe everything. And to my wife Renia M.T., whose patience, tolerance and understanding made this work possible.

ABOUT THE AUTHOR

Andrze J. Wortman, received highest undergraduate scholarship and Highest Honors with B.S. in mechanical engineering from the University of California, Berkeley. At the same institution he held the highest graduate scholarship for three years and got his M.S. in aeronautical science while pursuing advanced studies. He received his Ph.D. from the School of Engineering and Applied Science University of California, Los Angeles in heat and mass transfer. During 1971-1972 Dr. Wortman represented the U.S. Academy of Sciences in an exchange program with Poland where he has subsequently returned in similar capacity. Dr. Wortman has been under contract and has served as a consultant to major aerospace companies and many agencies of the U.S. Government. His publications include approximately 50 papers in the open literature and more than 200 company reports. He is a member of Sigma Xi, Pi Tau Sigma, Tau Beta Pi and is currently teaching Mechanical Engineering at California State University, Fullerton.

CONTENTS

PREFACE

The fossil fuel crises of the last decade stimulated a revival of interest in renewable energy sources. Considerable attention has focused on wind power, not only because it is an attractive energy source, but also because within the living memory of a significant fraction of the U.S. population very large numbers of small windmills could be seen in rural areas.

The perceived renewed need for wind energy systems stirred the enthusiasm and imagination of designers and inventors who in a short time produced a vast array of concepts. Many of the ideas which were not funded within the U.S. Government wind energy program were evaluated, together with all other submitted non-nuclear energy-related concepts, by the Office of Energy-Related Inventions of the National Bureau of Standards. It has been my pleasure and privilege to assist this Office in the evaluation of inventions involving fluid mechanics and thermodynamics, among which I have had to deal with many wind energy system ideas. The Office is headed by Mr. George P. Lewett whose staff of dedicated specialists handles vast numbers of submittals in all areas of science and engineering. In our efforts to provide evaluations and assistance to inventors, I have had the most pleasant and rewarding professional and personal contacts with Mr. Lewett, Mr. Tom Coultas, Dr. K.M. Tu, Mr. Howard Robb and other members of the staff too numerous to list here.

Under the sponsorship and guidance of the Office of Energy-Related Inventions, with Mr. Tom Coultas as the Technical Monitor, the analyses, technical data, and general facts on windmills which were accumulated in the course of many evaluations were organized and assembled in a report, Guidelines for the Evaluation of Wind Turbine Generating Systems. Subsequent evaluations and further study led to the expansion of the Guidelines into the present book. The range of topics and the emphases on certain problems reflect the scope and frequency of the evaluated inventions. The book extends the Guidelines as an aid to quick assessment of the technical and economic feasibility of a broad range of wind machines.

The book does not attempt to present a general introduction or an exhaustive treatment of the field of wind turbines. Instead, outlines of the current analytical techniques and engineering procedures are presented together with some fundamental derivations and references for more detailed study. Some of the theoretical treatments may be of interest to aerodynamicists, but the main goal is to present the techniques of wind turbine analysis in such a form that wind energy enthusiasts without any formal background in applied aerodynamics will be able to perform engineering estimates of current and new concepts in wind energy systems. For this purpose, a large part of the data is presented in graphical form. Methods for estimating the cost of energy derived from wind energy systems are also presented so that realistic feasibility studies can be performed.

My special expression of gratitude goes to the Office of Energy-Related Inventions, and to Mr. Tom Coultas in particular, without whose support and guidance this book could not have been written. As in countless similar circumstances, Gayl A. Brinlee performed herculean tasks to transform a mass of semi-legible notes and sketches into an engineering publication.

CHAPTER 1

INTRODUCTION

Wind is an old established source of energy which is clean, renewable, abundant and, in principle, accessible at no cost. Forces exerted by the wind are readily apparent and thus efforts to harness wind energy appeared very early in the history of mankind. Recent increases in the cost of fossil fuels which displaced wind power in the last century have revived the interest in modern exploitation of this form of solar energy. Numerous wind energy projects and enterprises enjoying private and government support have appeared on the scene with a bewildering array of concepts and claims. The explosive growth of this field has unfortunately led to the undertaking of many ill-conceived ventures whose resounding failures have placed the better programs in awkward defensive positions. The alacrity of individual and corporate entrants of uncertain credentials into the wind energy field was principally due to excessive influx of government funds which were appropriated on the basis of energy availability estimates derived from considerations of available land and average wind speeds. On the one hand excessive claims for wind energy have spawned numerous "energy" enterprises which attempt to market under-engineered windmills to ill-informed clients. On the other hand, the ample government funding has promoted the development of high technology systems of dubious technical and economic merit. Economic studies justifying large systems are usually based on production runs of thousands of multi-megawatt systems with total disregard of the problems of finding such a large number of suitable sites and the tens of billions of dollars for construction. The latest restructuring of the US energy policy should lead to a more deliberate, measured pace of development of the huge potential of wind energy, with logical engineering advances based on existing wind turbine systems.

HISTORICAL NOTES

Earliest windmills were drag-type machines, or panemones, in which the aerodynamic forces due to the wind pushed arms with simple flat surfaces. Net torque was generated by exposing the arms retreating from the wind and shielding the advancing arms by a barrier, or an enclosure. There are indications that crude versions of such machines with vertical axes may have existed as early as about 2000 B.C. in Babylonia and China. Traces of semi-enclosed vertical axis paddle-type windmills have been found in ancient Persia. Over two thousand years ago, Hero of Alexandria described a small horizontal axis windmill with sails as aerodynamic surfaces. Introduction of windmills to Europe is variously ascribed to

Arabs during their conquest of the Iberian peninsula, or to the Crusaders returning from the Near East a few centuries later. It is certain that by the twelfth century the use of windmills in Europe was established. Improvements in design and construction led to the appearance of the efficient English and Dutch windmills which were used for grinding, pumping and operation of machinery. In the nineteenth century the small American farm-type multi-bladed windmill was so popular that hundreds of thousands of machines were produced. The demand for electricity in the rural areas of the United States in the first half of the twentieth century was largely satisfied by the excellent Jacobs windmill. With the rural electrification program, the demand for the small windmills disappeared and only a few examples remain as testimony to their effectiveness and durability. A small number of manufacturers managed to survive until the revival of interest in wind power in the last ten years, and these offer well developed small units. For more historical notes the excellent treatment of Golding (1) should be consulted. The outlines presented by Eldridge (2) and Hunt (3) are also recommended.

LIMITS ON WINDMILL POWER

Some of the more extravagant claims for wind power and the subsequent disappointments of the users stem from the general lack of appreciation for the very low energy content in the wind and the sharp limitation on the fraction of energy which can be extracted. The common practice of rating turbines at very high wind speeds which occur during a very small fraction of the time has created considerable confusion among potential users. The average wind speed in the United States is about 10 mph. Because of the cubic variation of windmill power with wind speed, a wind turbine rated at 2 kW at 20 mph is more likely to show about 0.25 kW during a significant portion of the time.

The absolute upper limit on the amount of energy which can be extracted from the wind is set by the condition that at some distance downstream of the turbine, the pressure in the stream which has gone through the turbine must equilibrate with the ambient pressure. This is an elementary condition which must hold for all types of turbines and includes any augmenting devices, shrouds or diffusers. The upper limit on the extraction of power is generally referred to as the Betz limit although there are now indications that Lanchester may have discovered it first (Bergey (4)). Since the upper limit is set by fundamental laws and is therefore independent of the details of the turbine, an analysis developed by Rankine may be followed in replacing the turbine by a so called Froude actuating disk. In this analysis, the turbine, or propeller, are replaced by a mathematical surface across which the pressure is discontinuous, but velocity is constant. The system being analyzed is represented schematically in Figure 1.

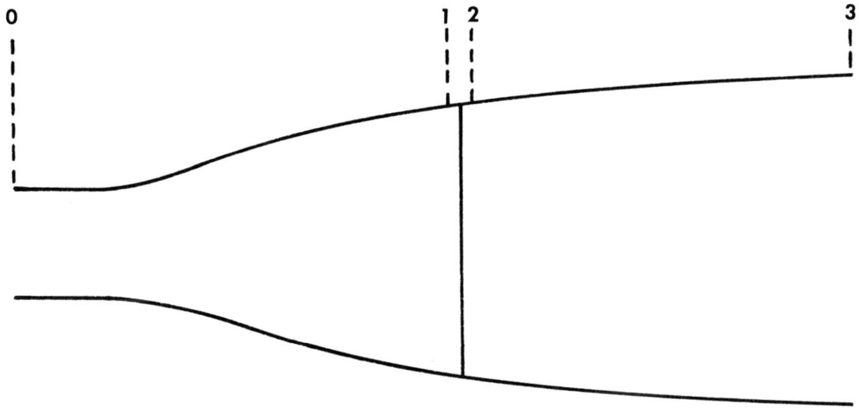

Figure 1. Actuator disk analysis of wind turbines

The four states of the flow are represented by subscripts 0, 1, 2, 3. The Bernoulli function between states 0 and 1 is constant as it also is between states 2 and 3. Thus, with p, v and ρ being the pressure velocity and density respectively

$$p_0 + \rho v_0^2/2 = p_1 + \rho v_1^2/2 \tag{1}$$

$$p_2 + \rho v_2^2/2 = p_3 + \rho v_3^2/2 \tag{2}$$

With the conditions $p_0 = p_3$ and $v_1 = v_2$ and definition of a

$$v_3 = v_0(1-2a) \tag{3}$$

the pressure across the disk is found to be

$$p_1 - p_2 = 2\rho v_0^2 a(1-a) \tag{4}$$

When the thrust on the disk is equated to the change in the impulse function

$$I = pA + \dot{m}v \qquad (5)$$

between states 0 and 3 it is found that

$$v_1 = v_0(1-a) \qquad (6)$$

This is the well known result in propeller theory that half the overall velocity change occurs before the actuator disk location. Rate of work done by the fluid i.e., the power extracted, P, is the thrust times the velocity.

$$P = \left(\frac{\rho v_0^3 A}{2}\right) 4a(1-a)^2 \qquad (7)$$

Power coefficient, C_p, is defined as the ratio of the power extracted to the power in the stream which is given by the terms in the brackets. Thus

$$C_p = 4a(1-a)^2 \qquad (8)$$

The maximum value of C_p is achieved at a = 1/3 and is equal to 16/27. It should also be noted that the flow through the disk corresponds to a stream tube of 2/3 the disk area and expands to a stream tube of twice the disk area.

The upper limit on efficiency of 16/27 cannot be reached because of aerodynamic losses. Any design claiming to approach this limit should be viewed with some suspicion. The power P which could be extracted by an ideal turbine of area A would therefore be given by

$$P = \frac{16}{27}\left(\frac{\rho v_0^3 A}{2}\right) \qquad (9)$$

A simple form of this relation for sea level operation gives the power P in kW as

$$P_{ideal} \cong \frac{1}{4}\left(\frac{V}{10}\right)^3\left(\frac{D}{10}\right)^2 \qquad (10)$$

with V in mph and D in ft. A relatively small real turbine might have an average efficiency e which is about 40% of the ideal value so that for quick estimates the relation becomes

$$P_{real} \approx \frac{1}{10}\left(\frac{V}{10}\right)^3\left(\frac{D}{10}\right)^2 \qquad (11)$$

A 10 ft diameter windmill which might be rated at 0.8 kW at 20 mph will produce only 0.1 kW at the average wind speed of 10 mph. The calculation for an annual energy output is even more discouraging since windmills cannot use the high end of the wind spectrum because of structural and electrical limitations. The low end of the wind speed spectrum is also useless since power varies as the cube of the wind speed. The annual energy output corresponds therefore to some fraction of the power at the average wind speed and the total time. If 6000 hours of operation at the equivalent of an average wind speed is assumed then the annual output of the windmill considered above would be 600 kWhr, or $45 in the value of electricity. At current prices the above turbine would cost about $3000 so that the return on investment would be very poor. On the other hand, if the average wind speed was 20 mph then the $280 annual return on investment would be much more reasonable.

CONCLUDING REMARKS

This book is a major expansion of "Guidelines for Evaluating Wind Turbine Generating Systems" (5) which was developed under contract to the National Bureau of Standards as an aid in the evaluation of wind turbine system concepts which were being submitted to the Office of Energy-Related Inventions. Some of the material presented here requires some knowledge of applied aerodynamics, but most of the presentation is aimed at the level of comprehension of wind energy enthusiasts with some technical background. Application of the engineering analyses and procedures presented here will develop a rational approach to the exploitation of the full potential of wind energy.

REFERENCES

1. Golding, E. W., The Generation of Electricity by Wind Power, E. & F. N., Spon. Ltd., London, U.K., Halstead Press, N.Y., 1978.
2. Eldridge, F. R., "Wind Machines," Report NSF-RA-N-75-051, National Science Foundation, Washington, D.C., Oct. 1975.
3. Hunt, V.D., Windpower, Van Nostrand Reinhold, New York, 1981.
4. Bergey, K. H., "The Lanchester-Betz Limit," AIAA Journal of Energy, Vol. 3, No. 6, Nov-Dec., 1979, pp. 382-384.
5. Wortman, A., "Guidelines for Evaluating Wind Turbine Generating Systems," Report AWD-R-81-0201, AWD Inc., Santa Monica, CA, Feb. 1981.

CHAPTER 2

EXTRACTION OF ENERGY FROM THE WIND

Wind energy has been exploited since the early stages of the development of civilizations to propel water craft and to drive machinery. Land based applications of wind energy began to wane with the appearance of practical steam engines which offered greater power and availability, rather than dependence on the vagaries of winds. Marine systems employing sails lasted well into this century, culminating with the Flettner rotors (1) which derived propulsion forces from the Magnus effect on rotating cylinders. Small scale wind turbine applications on forms could not survive the competition of inexpensive electricity which became available in rural areas during the major electrification programs. At present, the contribution of wind energy to the total energy utilization is negligible in all sectors of the economy. Interest in wind energy revived in the last decade when the cost of petroleum fuels increased by an order of magnitude. Despite the limited availability of suitable sites with high annual wind energy potential, and the uncertainties in wind occurrences, speeds and directions, it appears that wind energy systems will be increasing their contribution to the reduction of the dependence on fossil fuels.

APPLICATIONS OF WIND POWER

The energy delivered by a wind turbine is in the thermodynamic sense the highest form of energy since it is completely available and can be converted totally into any other form of useful energy. Thus, in principle, the captured wind energy can be applied to any imaginable situation. Actually, the applications are somewhat limited because of the variability of the wind turbine output with time and the overall cost of the delivered energy. The principal applications are outlined below.

Electrical Energy

The wind turbine drives an electrical generator whose output is either stored in batteries or linked with the main power grid. Linked systems require synchronization with the grid frequency which means that either the wind turbine must be operated at a constant angular rate or it must drive a variable input speed constant frequency output generator. Such machines are available in power output ranges up to 50-60 kW but are presently prohibitively expensive. Storage of electrical energy in batteries is expensive because of the cost of the batteries and the output equipment which is needed to convert direct current to the more commonly used alternating current. At the same time the efficiency of storage and retrieval of energy is only about 75% so that the overall cost of wind generated stored electrical energy is very high.

Chemical Energy

Wind power may be used to produce hydrogen and oxygen which can then be used as a clean burning fuel. Since the thermal efficiency of heat engines is quite low the overall efficiency of such a system is rather disappointing. Although such systems and their variants employing fuel cells have been suggested, the practicality of this approach remains to be demonstrated.

Thermal Energy

Wind power may be used to drive heat pumps, or to compress air for use in some thermodynamic cycles. The latter is quite frequently suggested despite its inherent overall inefficiency due to the pumping losses and the thermal efficiency of the heat engine. It is generally assumed that the storage tank will be insulated so that no heat will be lost from the compressed air to the surroundings. In an actual system, it is unlikely that the compressed air will be used immediately and it is necessary to consider the worst possible case of the cooling of the stored air to the ambient temperature. The efficiency of storage and retrieval is then readily calculated for the case of ideal compression and discharge of air.

When air is compressed isentropically from pressure p_1 to p_2 and the ratio of specific heats is γ the temperature ratio is given by

$$T_2 / T_1 = (p_2 / p_1)^{\frac{\gamma - 1}{\gamma}} \tag{1}$$

In a flow process the work done on the compressed gas is the enthalpy change H which is given by

$$H = mc_p(T_2 - T_1) \tag{2}$$

with m being the mass of the gas, and c_p the specific heat at constant pressure. The air will be stored in a tank where its temperature will rise because of flow work to T_3 which is given by

$$T_3 = \gamma T_2 \qquad (3)$$

and the pressure p_3 will equal to p_2. Assuming total equilibration of the temperature with the ambient at T_1 the pressure will fall to a value p_4 which is given by

$$p_4 / p_1 = \gamma^{-1} (p_2 / p_1)^{1/\gamma} \qquad (4)$$

When tanks are discharged the process is usually quite rapid and may be assumed to be adiabatic. The remaining air in the tank will expand isentropically to temperature T_1 with

$$T_t / T_1 = (p_1 / p_4)^{\frac{\gamma-1}{\gamma}} = \left[\gamma (p_1 / p_2)^{1/\gamma}\right]^{\frac{\gamma-1}{\gamma}} \qquad (5)$$

and the remaining mass fraction, m_t, in the tank will be

$$m_t / m = \rho_t / \rho = (p_1 / p_4)^{1/\gamma} \qquad (6)$$

The enthalpy of the discharged gas will be

$$h = m c_v T_1 \left[1 - \gamma (p_2 / p_1)^{-1/\gamma}\right] \qquad (7)$$

The efficiency of storage and retrieval, e, is h/H which may be expressed as

$$e = \frac{1}{\gamma} \left[1 - \gamma (p_2 / p_1)^{-1/\gamma}\right] \Big/ \left[(p_2 / p_1)^{\frac{\gamma-1}{\gamma}} - 1\right] \qquad (8)$$

Calculated values of e for air with $\gamma = 1.4$ are shown in Figure 1 for a wide range of pressure ratios. The moderate pressure ratios at which efficiency is relatively high are too low for efficient expansion engines so that the economics of wind power generated compressed air storage are somewhat doubtful.

Potential Energy

Wind turbines pump water to reservoirs which then run fluid machinery. This is an attractive concept because the overall efficiency of storage and retrieval is quite high and the variability of the power output of the wind turbines is inconsequential.

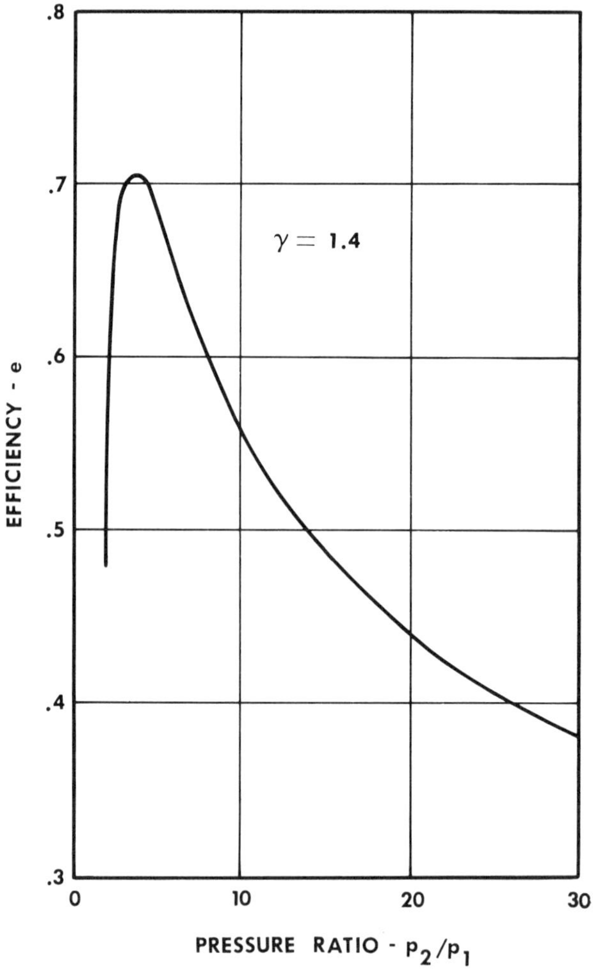

Figure 1. Efficiency of storage and retrieval of compressed air

Pumping and Irrigation

Wind power is used to deliver water for agricultural purposes. If sufficiently windy sites can be found then variations in power output over a period of hours or even days are not important and relatively inexpensive small windmills could easily prove to be economically attractive.

Heating

Wind turbine output is expected to generate heat through either solid or liquid dissipation devices. This is a fundamentally unsound idea since it converts the high grade mechanical energy of the windmill into the lowest grade of energy, heat.

Concluding Remarks

The above list is by no means exhaustive, but is intended to indicate the range of possible applications of wind power. Until it is demonstrated that the costs of very large wind turbines producing electricity as part of the main power grid can be reduced to an acceptable level, it would seem that relatively small, simple machines of 20-50 kW output, used to pump water are economically the most attractive.

SITING AND WIND CHARACTERISTICS

Economic considerations dictate the necessity of locating wind turbines on windy sites with high total annual energy potential. Distribution of wind speed in the continental United States is obtained from several hundred monitoring stations by the National Climatic Center in Asheville, North Carolina. Because of the sparseness of the data only crude contours of energy availability can be obtained. Eldridge (2) shows such contours for the world. The most windy areas are the west coasts of South America and north Africa, almost all the coast of Greenland and several smaller areas with available annual energy of over 5,000 kWh/kW of rated wind turbine output. The coastal regions of the United States are shown to have available energies of 2250-3750 kWh/kW, and the inland areas show availability of only 750-2250 kWh/kW. These, of course, are only gross approximations which may be used for rough indications of available sites in different regions. Local conditions and topography influence wind characteristics considerably, and reliable information can be obtained only from detailed long term measurements.

Selection of sites for wind power systems is influenced strongly by engineering, environmental, legal and practical considerations. Here, only the first will be considered since the others involve factors which are peculiar to the particular set of circumstances for each project.

Simple rules for selection of location follow from elementary aerodynamic considerations:

a. Locate on mildly sloping hills

The wind will be accelerated by the hill and an augmentation effect can be realized

b. Locate in canyons with prevalent winds

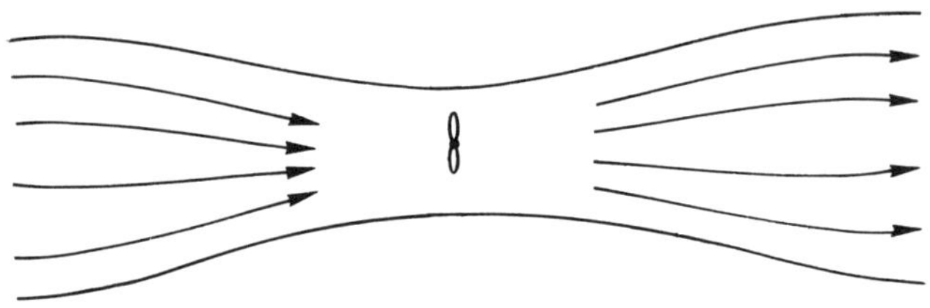

Canyons will accelerate the flow and maintain a constant direction

c. Avoid location on bluffs or near the edges of tall buildings

Flow separates from the leading edge and forms a region of recirculation which is extremely turbulent.

The above are general principles which should be followed, but in general, it is necessary to determine the optimum location with an integrating anemometer. Even then the information is incomplete because the total energy which can be extracted over a period of time will be determined by the variations of the wind turbine performance with wind conditions. An indication of the availability of energy in the winds around the world is given by Eldridge (2).

At any particular site, the wind velocity will be fluctuating in both magnitude and directions. Gusts may range up to 50% of the mean value and have periods of a fraction of a second to several seconds. In a period of several hours the magnitude of the mean speed may vary by 50% from the mean for the period. Thus, characterization of the wind at any site is extremely difficult and considerable experimental effort is necessary before any significant commitment of a design or financial effort is advised. Because of the enormous variations in wind characteristics with time and location, the mathematical models commonly employed are not in general agreement on the distribution of wind velocity in time, direction and space.

ENVIRONMENTAL CONSIDERATIONS

The cost of units of power produced by windmills decreases with increasing size. Because of the low power density of the wind the minimum size of windmills represents a rather large system, since a net output of 1 kW in a 19 mph wind requires a 12 ft. diameter blade.

The principal environmental concerns are:

a. Appearance of large, and rather ungainly structures. This is particularly true of wind turbines with shrouds, or augmentors.
b. Noise generated by the rotor and interference of flow with the supporting structure.
c. Interference with television reception due to both the rotating components and the fixed installation.
d. Hazards due to failure of moving components.
e. Vandalism by unlawful breaking and entering or gunfire which such unusual structures would almost certainly provoke.

FLOWS OVER HILLS

Calculations of flows over hills are extremely difficult because of the irregularity of the shape, the shear in the wind, and the possibility of thermodynamic effects induced by the pressure changes in the flows. However, Golding (3) suggests that an indication of the characteristics of the flow may be obtained from simple inviscid hydrodynamics analyses of flows over spheres and cylinders. In such analyses stream surfaces may be replaced by solid surfaces so that sphere flow stream surfaces will represent flow over a solitary spherical hill and cylinder flow stream surfaces will correspond to the flow over a long ridge. The hills do not have to be spheres or cylinders, but rather could be approximated by any particular stream surface.

For flow over a sphere of radius a, Streeter (4) gives the potential and stream functions, ϕ and ψ respectively, as

$$\phi = \frac{Ua^3}{2r^2}\cos\theta + Ur\cos\theta \tag{9a}$$

$$\psi = -\frac{Ua^3}{Ur}\sin^2\theta + \frac{Ur^2}{2}\sin^2\theta \tag{9b}$$

The radius r is measured from the origin of the coordinate system which coincides with the center of the sphere. The angle θ is measured from the axis of symmetry and the wind is towards the sphere in the $\theta = \pi$ direction. The tangential and radial components of velocity, v_θ and v_r respectively are given by

$$v_\theta = \frac{1}{r\sin\theta}\frac{\partial\psi}{\partial r} = -\frac{1}{r}\frac{\partial\phi}{\partial\theta}$$

$$= U\sin\theta\left[1 + \frac{1}{2}\left(\frac{a}{r}\right)^3\right] \tag{10a}$$

$$v_r = \frac{1}{r^2\sin\theta}\frac{\partial\psi}{\partial\theta} = -\frac{\partial\phi}{\partial r}$$

$$= -U\cos\theta\left[1 - \left(\frac{a}{r}\right)^3\right] \tag{10b}$$

Surfaces of constant flow speed, $q = \left(v^2 + v_r^2\right)^{\frac{1}{2}}$, normalized by the flow at infinity U are found from the solution of

$$\left(\frac{a}{r}\right)^{-6} + 4\left(\frac{a}{r}\right)^{-3}\frac{1 - 3\cos^2\theta}{1 + 3\cos^2\theta} = \frac{4\left(\frac{q}{U}\right)^2 - 1}{1 + 3\cos^2\theta} \tag{11}$$

which is quadratic in the cube of a/r, and can be solved using standard methods. Surfaces of constant dimensionless stream function ψ/Ua^2 and constant speed q/U are shown in Figure 2 which illustrates the limited extent of the flow field disturbed by the presence of a sphere.

A. Streamlines

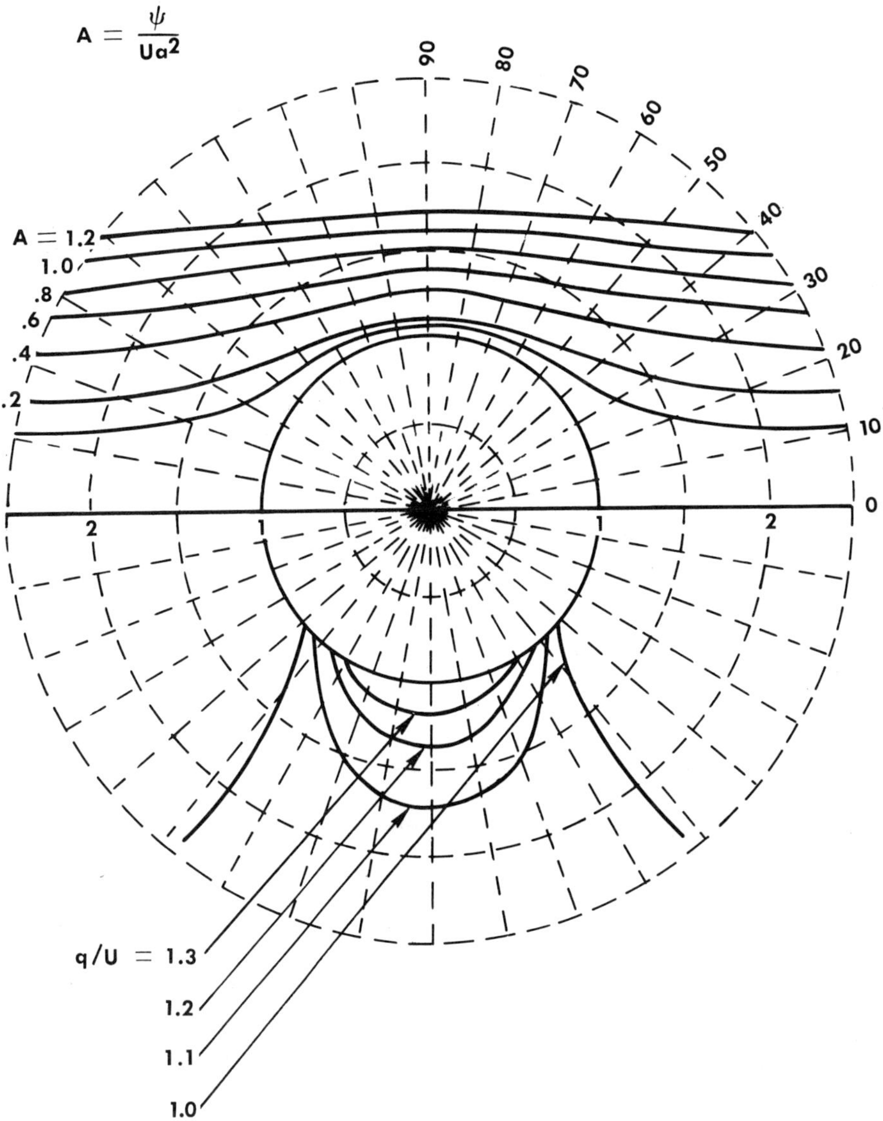

B. Isovelocity contours

Figure 2. Flow around a sphere

Solutions for flow over a cylinder are given by the potential and stream functions

$$\phi = U(r + a^2/r)\cos\theta \tag{12a}$$

$$\psi = U(r - a^2/r)\sin\theta \tag{12b}$$

with tangential and radial components of velocity given by

$$v_\theta = \frac{\partial\psi}{\partial r} = -\frac{1}{r}\frac{\partial\phi}{\partial\theta} = U\left(1 + \frac{a^2}{r^2}\right)\sin\theta \tag{13a}$$

$$v_r = -\frac{1}{r}\frac{\partial\psi}{\partial\theta} = -\frac{\partial\phi}{\partial r} = -U\left(1 - \frac{a^2}{r^2}\right)\cos\theta \tag{13b}$$

Streamlines defined by constant values of $\psi/Ua = A$ are calculated directly from the solution of Equation (12b) in the form

$$\frac{r}{a} = \frac{A}{2\sin\theta} + \left[1 + \left(\frac{A}{2\sin\theta}\right)^2\right]^{\frac{1}{2}} \tag{14}$$

Lines of constant speed are calculated with the solution of the equation

$$\left(\frac{r}{a}\right)^{-4} + 2\left(1 - 2\cos^2\theta\right)\left(\frac{r}{a}\right)^{-2} = \left(\frac{q}{U}\right)^2 - 1 \tag{15}$$

Calculated results for constant A and constant q/U are shown in Figure 3.

TYPES OF WIND TURBINES

Numerous concepts have been presented for wind turbines, but almost all of them fall into the two main classes of machines which extract power by employing lift of surfaces and machines which depend primarily on the drag of the rotating or translating elements. Further distinctions can be drawn between machines whose axes of rotation are parallel to the wind and those whose axes are normal to it. Even further differentiation is possible between machines whose elements rotate and those whose elements translate. Precise classification is further complicated by machines which combine some, or all, of the basic features. Additional problems are introduced by machines which employ some form of stationary or moving augmenting devices, inlet ducts, ejector shrouds,

A. Streamlines

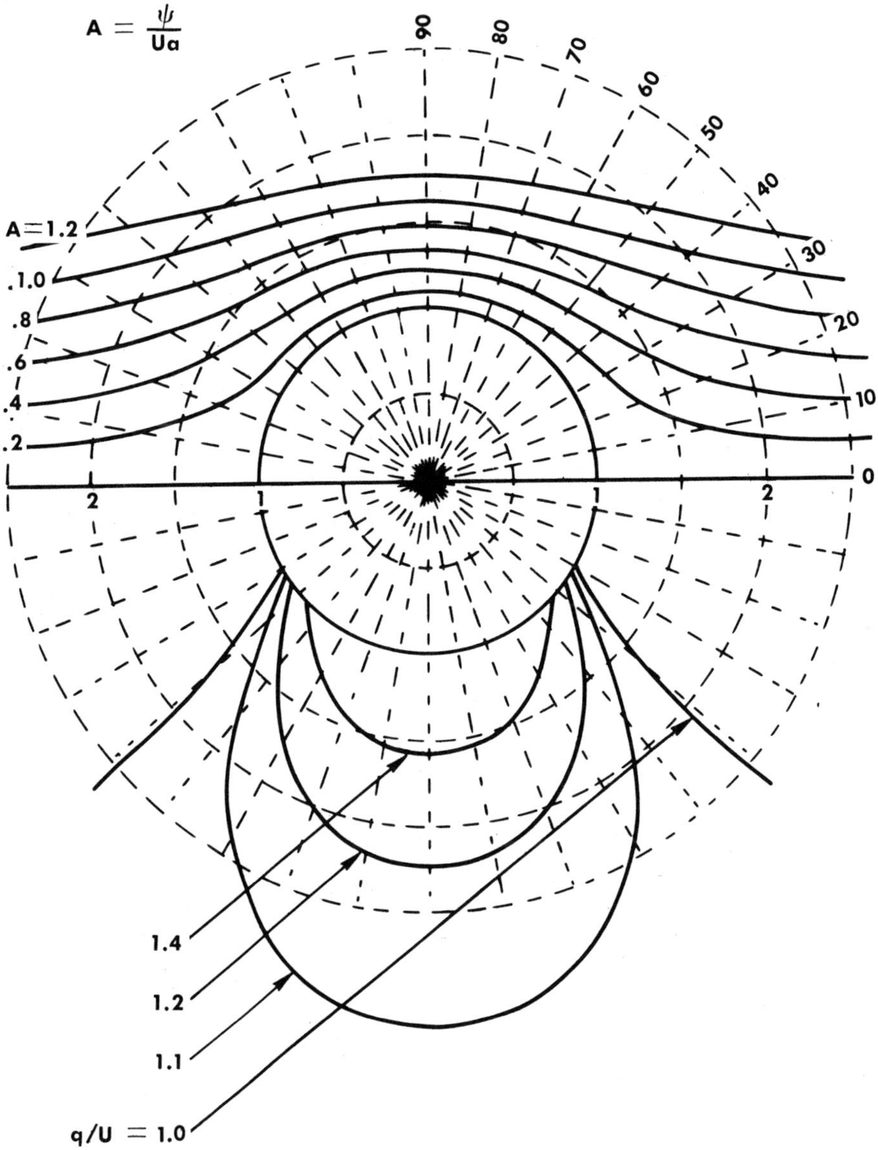

B. Isovelocity contours

Figure 3. Two-dimensional flow over a cylinder

diffusers and thermodynamic effects of solar or fossil fuel heating. Only the principal types of wind turbines will be mentioned here. For a comprehensive review of types of wind turbines the report by Eldridge (2) is recommended.

a. Axis aligned with the wind - lifting elements

Farm Type

Farm Type

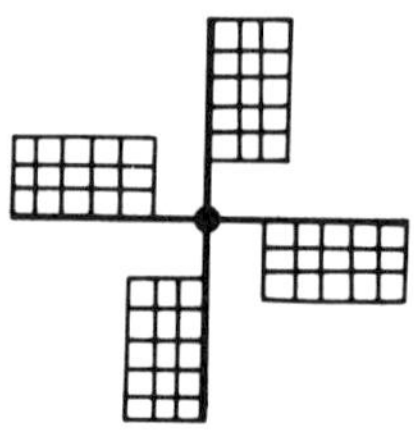

Dutch

Modern Propeller

Characteristics - Farm type and Dutch have excellent starting torque and power coefficients of 0.15 to 0.2

Propeller type have poor starting torque and power coefficients of 0.4 to 0.5

b. Axis normal to the flow - lifting elements

ϕ- Darrieus

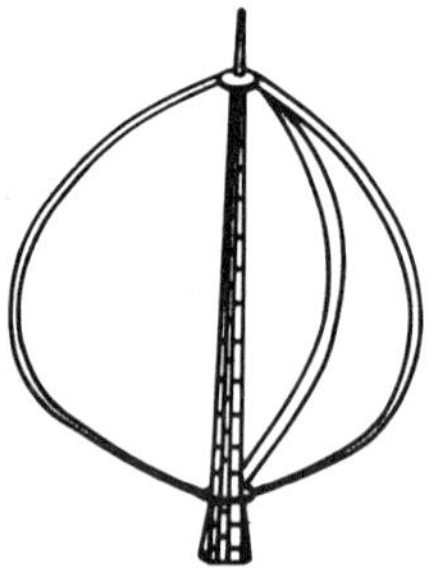

Δ- Darrieus

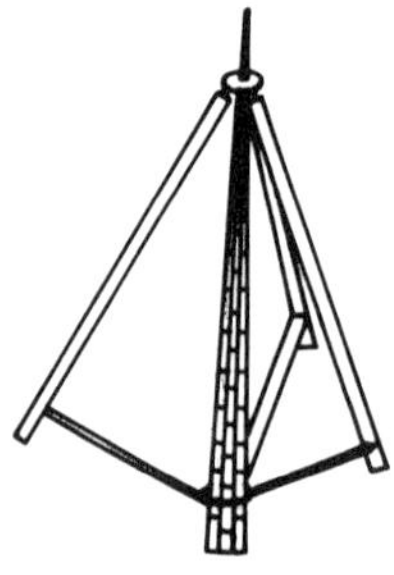

Giro-mill

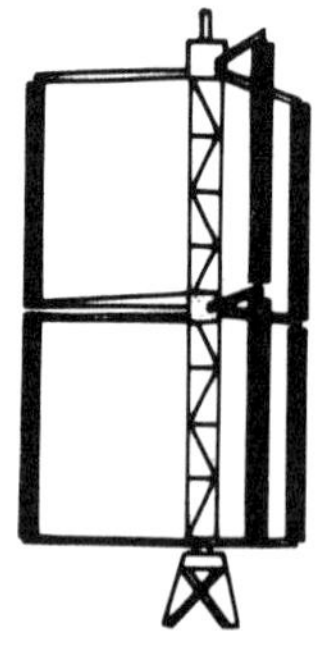

Characteristics - Very poor starting torque and power coefficients of 0.4 to 0.5

c. Axis normal to the flow - drag elements

Savonius   Split Savonius   Shield

Characteristics - Excellent starting torque and power coefficients of 0.15 to 0.3

Because of the enormous variation in concepts for augmenting the output of wind turbines, it is impractical to list even a representative sample here. In general, the augmenting concepts may be classified as follows:

a. Stationary surfaces - the flow is channelled into the active elements, but the main effect is to increase the capture area
b. Inlet ducts - a larger stream tube is captured
c. Ejector ducts - these mix the spent flow from the turbine with the wind and send it through a diffuser in the hope that the average energy of the flow will be sufficient for diffusion to the free stream pressure
d. Diffuser ducts - these are attached downstream of the turbine in the hope that the spent flow will diffuse to ambient pressure more efficiently than it would on its own

All of the above ideas share the hope that by adding such devices a more efficient operation of the turbine will be realized. Actually, fundamental considerations outlined in the preceding chapter demonstrate that on the basis of the projected area of the device, the upper limit on power extraction is fixed by laws of physics. If the above devices are viewed as stationary but possibly orientable structures which may realize a certain power output with a smaller rotating element then some justification for their consideration can be admitted. On the other hand, a 10 ft. radius turbine with a 50 ft radius, 285 ft long diffuser of 8° divergence angle is not likely to be competitive with a 50 ft radius simple propeller. If nothing else than the problems of aligning such an enormous device with the wind, and the visual impact will probably eliminate various forms of shrouds from any serious considerations. Efficiency curves of turbines are shown in Figure 4 which is taken from Eldridge (2).

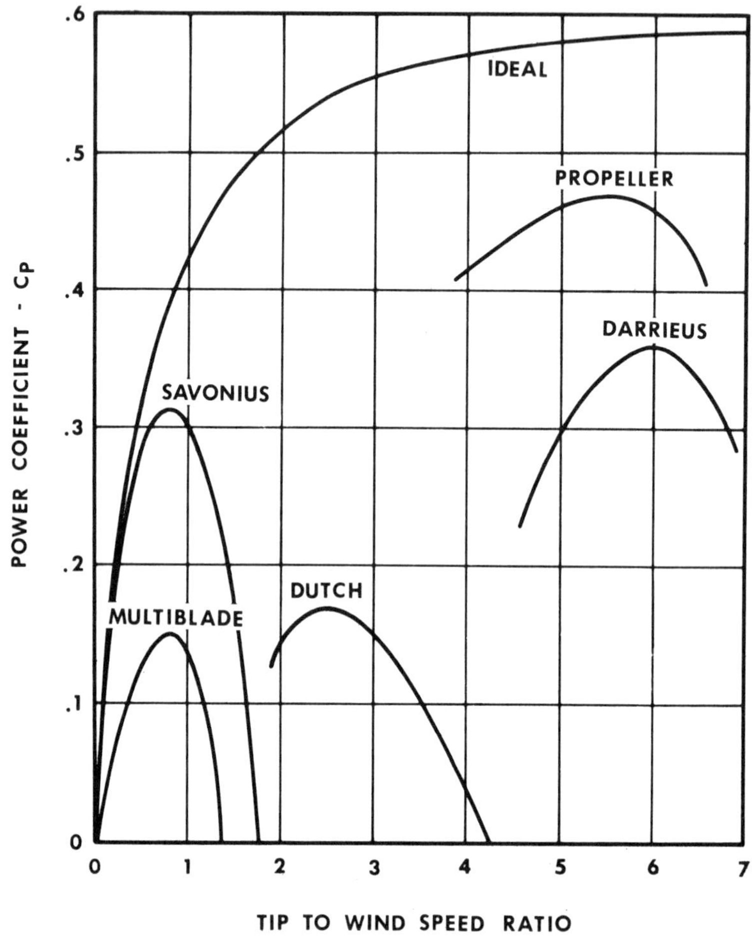

Figure 4. Representative power coefficients of wind machines

IDEAL WINDMILL PERFORMANCE

In this section attention will be focused on the performance of propeller-type windmills which will be analyzed under the assumption that a radial strip of width dr, at a radius r is not influenced by the neighboring elements. Also, advantage is taken of the previously derived result that half of the total reduction of the velocity of the wind is realized at the propeller disk. The analysis presented here follows generally along that of Glauert (5). If the axial velocity through the disk is v', the induced angular velocity is ω and the wind velocity is V then the momentum equations for the drag D and torque T are

$$\frac{dD}{dr} = 4\pi r \rho v'(V - v') \tag{16a}$$

$$\frac{dT}{dr} = 2\pi r^3 \rho v' \omega \tag{16b}$$

With the definitions

$$v' = V(1 - a)$$

$$\omega = 2\Omega a'$$

where Ω is the angular rate of the blades, the above equations become

$$\frac{dD}{dr} = 4\pi r \rho V^2 (1 - a)a \tag{17a}$$

$$\frac{dT}{dr} = 4\pi r^3 \rho V \Omega (1 - a)a' \tag{17b}$$

The energy balance for the stream is given by

$$(1 + a')\Omega \frac{dT}{dr} = (1 - a)V \frac{dD}{dr} \tag{18a}$$

or upon substitution of Equation 17b

$$(1 + a')a'x^2 = (1 - a)a \tag{18b}$$

where

$$x = r\Omega/V$$

The local element of power is $dP = \Omega dT$ and the element of power coefficient is given by

$$dC_p = \frac{dP}{\pi R^2 \rho V^3/2} = \frac{8}{X^2}(1 - a)a'x^3 dx \tag{19}$$

where $X = R\Omega/V$, the tip speed ratio. The maximum of dC_p is found by maximizing $(1 - a)a'$ which leads to

$$(1 - a)\frac{da'}{da} = a' \tag{20}$$

Differentiation of Equation 18b with respect to a yields

$$(1 + 2a')x^2 \frac{da'}{da} = (1 - 2a) \tag{21}$$

which upon substitution of Equation 20 becomes

$$(1 + 2a')a'x^2 = (1 - a)(1 - 2a) \tag{22}$$

When this is combined with Equation 18b) a relation between a and a' is obtained in the form

$$a' = \frac{1 - 3a}{4a - 1} \tag{23}$$

which upon substitution into Equation 18b or Equation 22 yields

$$a'x^2 = (1 - a)(4a - 1) \tag{24}$$

Sufficient information now exists to relate the performance parameters a, a', x and $X^2 dC_p/dx$ which is done in Figure 5.

The above relations may be used to determine propeller characteristics for ideal, zero drag blades. The relative wind vectors at a blade element are shown in Figure 6. This time the elemental drag and torque relations are written as

$$\frac{dD}{dr} = \frac{1}{2} Nc\rho W^2 C_L \cos\phi \tag{25a}$$

$$\frac{dT}{dr} = \frac{1}{2} Nc\rho W^2 C_L \sin\phi \tag{25b}$$

with the relative flow angle ϕ defined by the relation

$$\tan\phi = \frac{1 - a}{1 + a'} \frac{V}{r\Omega}$$

Comparison of the above relations with the corresponding ones given in Equations 16a, 16b leads to the result

$$\frac{a}{1 - a} = \sigma C_L \cot\phi \csc\phi/4 \qquad (26)$$

$$\frac{a}{1 + a'} = \sigma C_L \sec\phi/4 \qquad (27)$$

where σ is the local solidity defined as

$$\sigma = Nc/2\pi r$$

with N being the number of blades and c the chord at radius r. Combination of Equations 26 and 27 and elimination of a' yields the relation

$$x\tan\phi = 1 - a\sec^2\phi \qquad (28)$$

The power developed by the element dr is $dP = \Omega dT$ which may be expressed in terms of the local power coefficient

$$c_p = \frac{dP/dr}{\pi r \rho V^3} = 4a(1 - a)x\tan\phi \qquad (29)$$

Following Stewart (6) optimum operation is determined by differentiation of Equation 29 with respect to ϕ, using Equation 28 to define the relation between a and ϕ. Optimum operation is found to correspond to:

$$a^{-1} = 2 + \sec\phi \qquad (30)$$

$$C_L = 4(1 - \cos\phi) \qquad (31)$$

$$x = \frac{(2\cos\phi - 1)(1 + \cos\phi)}{\sin\phi(1 + 2\cos\phi)} \qquad (32)$$

With these relations, the distribution of the blade loading parameter σC_L, relative flow angle ϕ, and the interference factor a with the local tangential speed ratio $x = r\Omega/V$ are determined for the optimum operation of an ideal windmill.

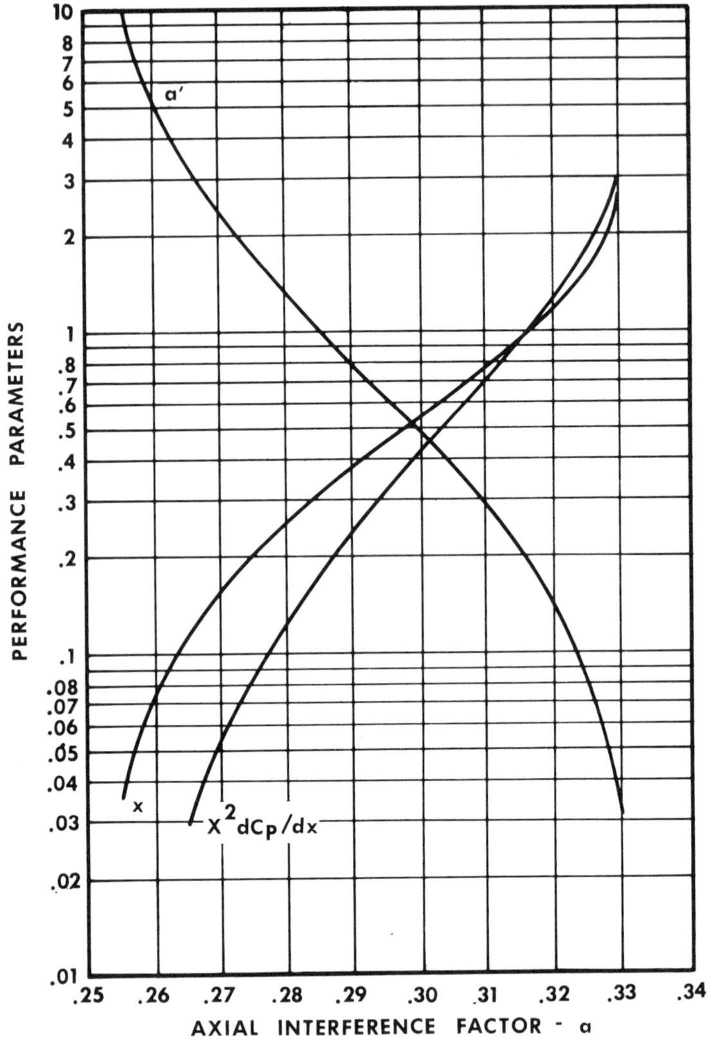

Figure 5. Performance parameters of an ideal windmill

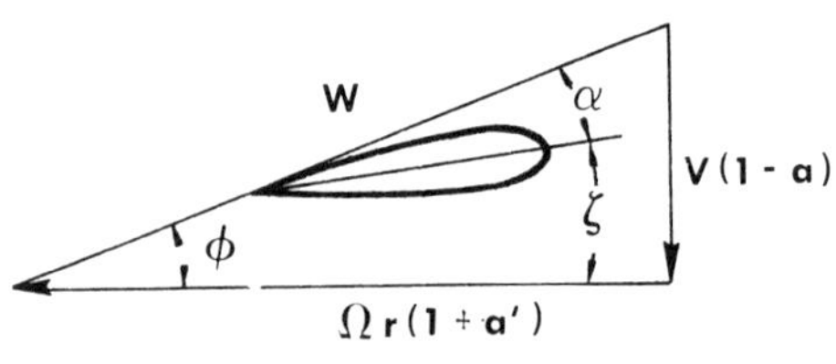

Figure 6. Relative wind triangle on a blade element

REFERENCES

1. Flettner, A. The Story of the Rotor, F.O. Willhoft, New York, 1926.

2. Eldridge, F. R., "Wind Machines," Report No. NSF-RA-N-75-051, National Science Foundation, Oct. 1975.

3. Golding, E. W., The Generation of Electricity by Wind Power, Halstead Press, New York, 1978.

4. Streeter, V. L., Fluid Dynamics, McGraw-Hill Book Co., New York, 1948.

5. Glauert, H., "Airplane Propellers," Aerodynamic Theory, W. F. Durand, Ed., J. Springer, Berlin, 1935, pp. 324-332.

6. Stewart, H. J., "Dual Optimum Aerodynamic Design for a Conventional Windmill," AIAA Journal, Vol. 14, No. 11, Nov. 1976, pp. 1524-1527.

CHAPTER 3

ENERGY IN THE WIND

The geographical distributions of wind intensities and durations which are indicated in currently available maps are only rather vague indications of the general locations for potential wind turbine sites. Some of the apparently high wind energy areas are in remote locations far removed from the potential energy consumption centers, and thus would be of minor commercial interest because of logistics problems and transmission costs. In order to present the potential energy extraction on a uniform basis it is customary to rate geographical areas in terms of the annual energy availability per unit power at some specified wind speed. The units employed are normally kilowatt hours per kilowatt of rated power. This is abviously a gross simplification since in many areas the nominal power in terms of average wind speeds is statistically somewhat misleading due to large diurnal and seasonal variations, and most wind machines would not be able to take advantage of the full range of wind speeds.

GEOGRAPHICAL WIND DISTRIBUTIONS

It is generally realized that wind speeds are influenced strongly by local topographical features, but it is generally possible to describe certain parts of the world as either high or low wind energy areas. The actual availability of wind energy is subject to some argument because of the lack of precise long term measurements. Golding (1) in his book, The Generation of Electricity by Wind Power presents large quantities of statistical data on winds, and points out that some of the earlier estimates of the available energy in the wind at the designated high energy sites may be too high by as much as 50%. In view of the present high cost of wind energy capture such data underscore the need for detailed local measurements prior to the undertaking of any wind energy projects. A rather detailed map of the distribution of wind energies around the world is given by Eldridge (2) in terms of the kWh/year per rated kW output for wind machines designed for nominal wind speeds of 25mph (40 km/h). In view of Golding's data some of the results given by Eldridge may be too optimistic, but it is interesting to note that several areas of the world such as southern Greenland, the southern tip of South America, and New Zealand appear to have available wind energy of over 5000 kWh/year per kW of rated power. In the United States both the East and West Coasts are shown to have available energy ratings of 2250-3750 kWh/year per kW. The central California region and the Great Plains are shown to have annual average wind speeds exceeding 18mph (29km/K), with a total power potential of several times the current United States needs.

LOCAL WIND DISTRIBUTIONS

It is extremely difficult to characterize winds, even with the availability of detailed local data because of the extreme transients in speed and direction which are exhibited at most sites. Very few localities exhibit prevalent winds which come from a particular direction. In general, winds tend to come from general geographical orientations such as East or North, with relatively few places where winds come from a particular quadrant, and considerably fewer places where a more precise definition of direction can be employed. Temporal variations on a scale of minutes are quite normal, and in most cases longer period variations with a time scale of hours are encountered. Golding (1) indicates that while some locations exhibit strong diurnal variations there are places such as St. Helena where the wind is essentially constant. Seasonal variations with a time scale of months are strongly influenced by the geographical location and must be determined in long term measurement programs. Even longer term variations are found from year to year, with some weak correlations with the sunspot cycles. These require historical data for periods of the order of decades, and are thus available for a very limited number of sites.

In view of the significant variability of the winds, it is not entirely clear whether it is worthwhile to design wind turbines for the highest possible efficiency at rated conditions, as is currently practiced, or whether less efficient designs with greater tolerance for variations in operating conditions might prove to be more cost effective. Wind turbine system studies require models of wind speeds which allow the assessment of the practical feasibility and commercial acceptance of the project. Detailed engineering studies must take into account such factors as gusts in order to design for adequate fatigue life and the time rate of change of the wind direction which determines the requirements for the usually expensive yaw control system in larger propeller wind machines.

The importance of yawing propeller turbines into the wind is easily demonstrated by estimating the power of a turbine with a fixed orientation relative to the power of a machine turning into the wind. Assuming a completely random distribution of wind directions, no power output when the wind is coming from the rear of a fixed orientation machines, and the variation of power with the cube of the velocity component normal to the disk the energy capture ratio R_E is given by

$$R_E = 2\int_0^{\pi/2} \cos^3\theta d\theta \Big/ 2\pi = \frac{2}{3\pi} \tag{1}$$

A fixed orientation turbine would capture only about 21% of the energy captured by a freely yawing turbine. Another interesting estimate is the effect of gusts on the power output of a turbine. Accurate definition of the gust component of the wind can only be obtained from precise measurements employing high frequency instrumentation. The response of the turbine to such rapid variations is extremely difficult to determine so that the problem represents a significant engineering challenge. However, it is possible to make simple estimates of the orders of magnitude of the quantities involved using the simplifying assumption of a harmonic variation of gusts with amplitude A and a period T. Again, assuming a cubic variation of power output with velocity, the average power relative to the nominal speed value, R_P, is found to be

$$R_P = T^{-1}\int_0^T \left[1 + A\sin(2\pi t/T)\right]^3 dt = \left(1 + \frac{3A^3}{2}\right) \tag{2}$$

With 20% gust amplitudes and the assumptions stated above the power would be 6% greater than the nominal value.

VERTICAL WIND SPEED GRADIENT

All land based systems operate in the terrestrial boundary layer in which occurs the transition from zero wind speed at the surface to the atmospheric wind speeds at high altitudes. This boundary layer extends to a height of several hundred meters and is influenced strongly by the surface conditions, insolation and humidity. Because of the large number of variables involved models of the velocity distribution employ numerous simplifications. The development presented here follows the general turbulent boundary layer analyses such as may be found in Schlichting's, Boundary Layer Theory (3), but is actually based on the excellent exposition of atmospheric boundary layers given by Kopcewicz (4).

Near the surface of the earth the equations of motion reduce to

$$\frac{\partial p}{\partial x} = \frac{\partial}{\partial z}\tau_{xz} \tag{3}$$

where x and z are the horizontal and vertical coordinates respectively, p is the pressure and τ_{xz} is the shear stress in the direction of x on a plane whose normal coincides with z. Near the surface p is independent of z and integration yields

$$\tau_{xz} = \tau_0 + z \frac{\partial p}{\partial x} \tag{4}$$

with τ_0 being the surface value of the shear stress. Near the surface the second term on the right hand side of the equation may be neglected. In terms of the Prandtl mixing length theory which is discussed in detail by Schlichting (3) the shear stress may be expressed as

$$\tau_{xz} = \rho \ell^2 \left(\frac{\partial u}{\partial z}\right)^2 \tag{5}$$

with ρ being the density, u the horizontal component of velocity and ℓ the mixing length. Combination of Equation 5 with the reduced form of Equation 4 leads to

$$\frac{\partial u}{\partial z} = \frac{1}{\ell}\sqrt{\frac{\tau_0}{\rho}} = \frac{u^*}{\ell} \tag{6}$$

where u* is the friction velocity. With the usual smooth surface assumption that ℓ = kz the above equation may be integrated directly to yield

$$u = \frac{u^*}{k} \ln z + C \tag{7}$$

with k being the von Karman constant which is equal to 0.4, and C = 5.5. In general, the surface of the earth cannot be considered to be smooth and the mixing length is taken as

$$\ell = k(z + z_0) \tag{8}$$

which leads to the relation

$$u = \frac{u^*}{k} \ln\left(\frac{z + z_0}{z_0}\right) \tag{9}$$

From this relation the turbulent eddy diffusion coefficient is found to be

$$\nu_t = k(z + z_0)u^* \tag{10}$$

If measurements indicate a velocity u_a at a height z_a then the eddy diffusion coefficient is given by

$$\nu_t = k^2(z + z_0)u_a \Big/ \ln\left(\frac{z_a + z_0}{z_0}\right) \tag{11}$$

In principle the earth's boundary layer could be calculated, provided the value of z_0 could be determined. Kopcewicz (4) gives z_0 = 0.001 cm for smooth ice, 0.1 cm for 1 cm grass, 0.7 - 2.3 cm for grass up to 10 cm (4") and 5 - 9 cm for grass up to 50 cm (20").

It is customary in meteorology to correlate wind speed measurements using the relation

$$u = u_1(z/z_1)^m \tag{12}$$

where m varies between 1/3 and 1/7, depending on local atmospheric conditions. The strong influence of the temperature gradient is shown in Figure 1 which is taken from Kopcewicz (4). A more direct illustration of the influence of temperature gradients on the value of the exponent m is shown in Figure 2 which is also taken from Reference 4.

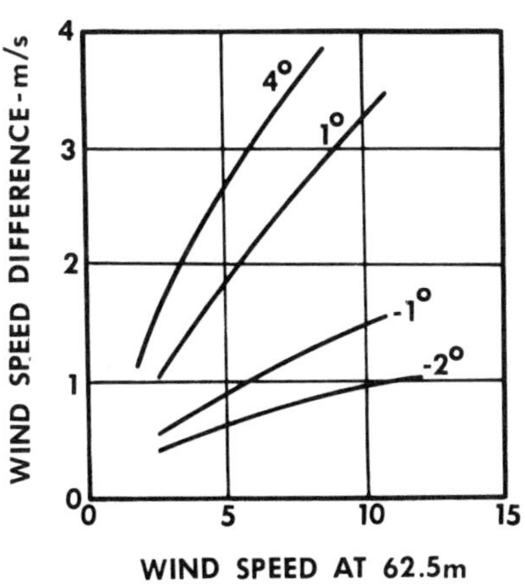

Figure 1. Dependence of the speed difference at heights of 62.5 and 15.2 m on vertical temperature gradients in °C/100m

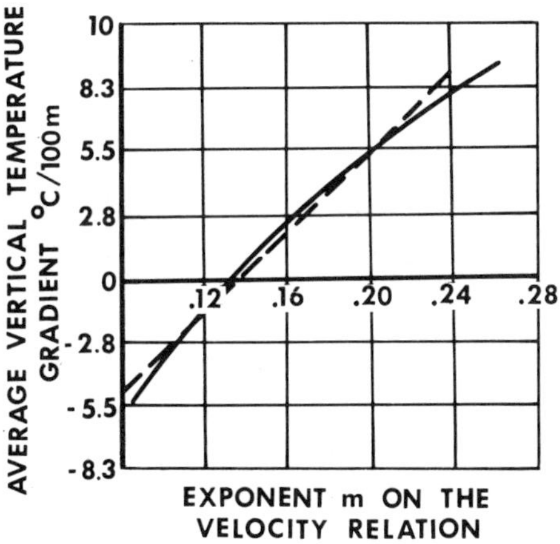

Figure 2. Dependence of the exponent m on the vertical temperature gradient

The exponent m may be expressed in terms of the logarithmic profile as

$$m = z/(z + z_0)\ln\left(\frac{z_0 + z}{z_0}\right) \tag{13}$$

which shows a dependence on the height z. The above discussion is intended to illustrate that simple relations for describing the vertical distribution of wind speeds must be treated with considerable caution. The influence of the velocity gradient on the performance of propeller wind turbines may be estimated using the power relation (Equation 12) cubic dependence of power on the wind speed and a simple blade element at radius r on an arm whose axis is at $z = h_0$. With the arm at angle to the horizontal, the instantaneous height of the element is

$$h = h_0 + r\sin\theta \tag{14}$$

The ratio of average power to the nominal power at h_0, R_h, is therefore given by

$$R_h = \frac{1}{2\pi}\int_0^{2\pi}\left(\frac{h}{h_0}\right)^{3m} d\theta = \frac{1}{2\pi}\int_0^{2\pi}\left(1 + \frac{r}{h_0}\sin\theta\right)^{3m} d\theta \tag{15}$$

This relation is easily integrated numerically for various values of the parameter r/h_0. The results, including the extreme value of $r/h_0 = 1$ are given in Table I.

Table I. Reduction of power output by wind speed gradient

m \ r/h_0	0.0	0.25	0.50	0.75	1.0
1/7	1.00	0.990	0.984	0.959	0.897
1/5	1.00	0.996	0.984	0.961	0.911

In general, the tips of the blades of a propeller turbine would not be permitted to operate very far below $r/h_0 = 0.5$ so that the influence of the wind speed gradient is seen to be small, and that of the details of the distribution negligible.

ANNUAL ENERGY CAPTURE

Feasibility of a wind turbine system is altimately judged on the basis of the return on the invested capital. It is therefore necessary to make estimates of annual energy output which can be translated into financial equivalents. Preceding sections indicated the difficulties associated with wind characterization so that any model for annual wind speed distributions must be viewed as an approximation. The most popular models are of the form

$$H = a\, e^{-b\lambda^c} \tag{16}$$

with H being the number of hours that the wind speed exceeds V with a mean value of V_a, a total number of hours in a year, b and c constants, and $\lambda = V/V_a$. In a request for proposal Solar Energy Research Institute, SERI (5) suggested the use of the following constants:

$a = 8766$ hrs $b = \pi/4.06$ $c = 2.27$

and average speed $V_a = 5.4$ m/s at a reference height $z_r = 9.1$ m. At the same time, the mean velocity profiles were specified as

$$V_a = V_r \ln(z/z_0)/\ln(z_r/z_0) \tag{17}$$

The surface roughness length z_0 for engineering applications was recommended as 0.05m (0.16 ft) which was stated to represent moderate terrains. Other terrains were characterized as very smooth with $z_0 = 0.0005$m, and $z_0 = 0.34$m for very rough areas.

With the customary assumption of the cubic dependence of power on wind speed, the energy dE extracted in time dH is

$$dE = -\left(C_P \frac{\rho}{2} V_a^3\, a\right) bc\lambda^{c+2}\, e^{-b\lambda^c}\, d\lambda \tag{18}$$

where C_P is the power coefficient. The ratio of the annual energy capture to that corresponding to the average wind speed which is in the parenthesis, R_A is known as the energy capture factor and is given by

$$R_A = bc \int_0^\infty \lambda^{c+2}\, e^{-b\lambda^c}\, d\lambda \tag{19}$$

The effect of varying the parameters b anc c on the duration H is shown in Figure 3. It should be noted that the SERI model is quite close to the Gaussian distribution which is given by b = 1, c = 2. The energy capture factors in Figure 4 are plotted against speed ratios since most wind machines could not take advantage of the full wind speed range. The curves with the unrealistically low values of c = 1 are included for comparison only. The curves for c equal to or greater than 2 tend to finite values as the wind speed ratio increases without limit. Actually, the limiting values are achieved at wind speed ratios of about 3.

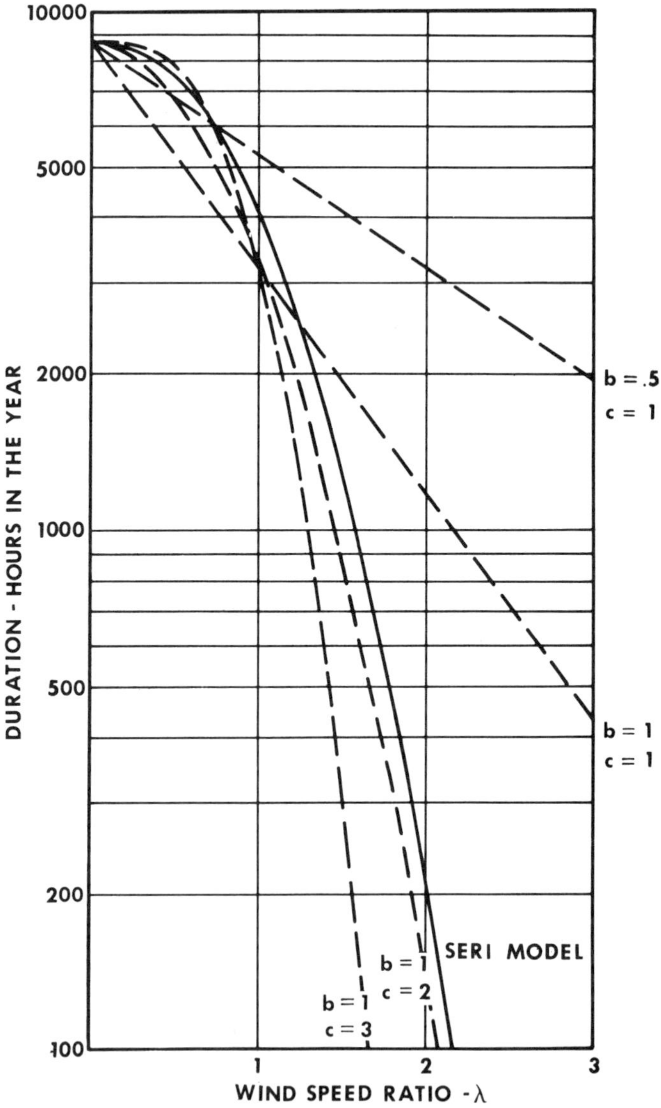

Figure 3. Duration of wind speed ratios exceeding given values

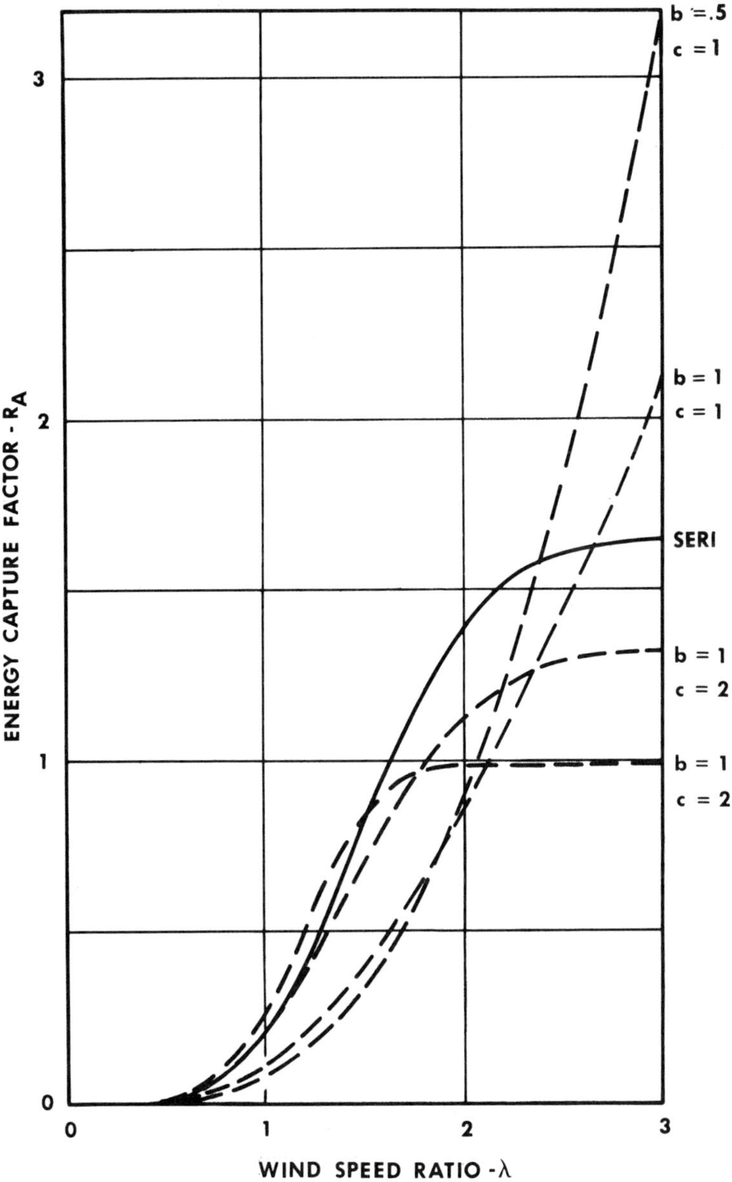

Figure 4. Influence of the wind duration equation parameters on the energy capture factor

REFERENCES

1. Golding, E. W., The Generation of Electricity by Wind Power, Halstead Press, New York, 1977, p. 265.

2. Eldridge, F. R., "Wind Machines," Report NSF-RA-N-75-051, National Science Foundation, 1975.

3. Schlichting, H., Boundary Layer Theory, Seventh Ed. McGraw-Hill Book Co., New York, 1979.

4. Kopcewicz, T., Fizyka Atmosfery, Part III, PWN, Warsaw, Poland, 1957.

5. Solar Energy Research Institute, Request for Proposal No. RH-9-8003, Generic Studies of Wind Energy Systems, Oct. 1978.

CHAPTER 4

PROPELLER TURBINES

The category of propeller turbines spans a very wide range of designs with the relatively low angular rate Dutch windmills at one end and the modern very high speed slender propellers at the other. Despite considerable effort, there are no exact theories for the operation of propeller turbines and it is necessary to rely on approximate models of the actual flow situation for determination of characteristics and predictions of performance. Probably the best engineering model is the one which combines the actuator disk analysis illustrated in Chapter 2 with aerodynamic analysis of an element of an airfoil. The vortex blade element analysis of Glauert (1) while introducing the parameter which includes the relative blade area is nevertheless essentially an infinite number of blades model which yields surprisingly good results. The analysis shown here follows along the general lines of the exposition of propeller theory presented by von Mises in his <u>Theory of Flight</u> (2).

The flow at an element of width dr located at radius r on a blade rotating at an angular rate Ω is shown in Figure 1

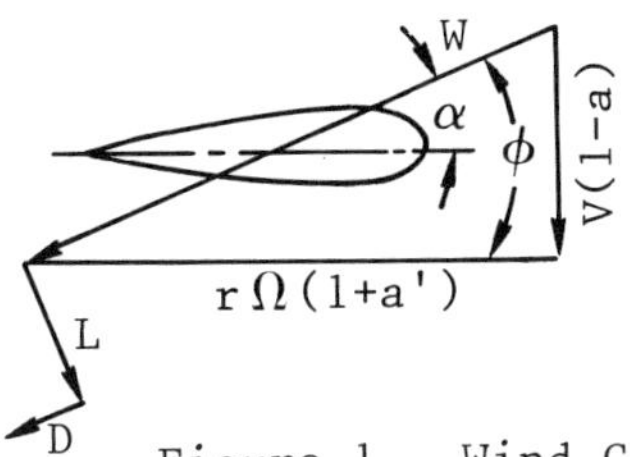

a = Axial interference factor
a'= Tangential interference factor
D = Drag
L = Lift
V = Wind speed
W = Relative wind speed

Figure 1. Wind Components and Aerodynamic Forces

The Bernoulli equation for a streamtube ahead of the propeller disk between a plane very far in front of the propeller, identified by V and p_0, and a plane immediately in front is written as

$$\frac{\rho}{2} V^2 + p_0 = \frac{\rho}{2}\left[V^2(1-a)^2 + w^2\right] + p \qquad (1)$$

The radial velocity w is relatively small and implicit in the above is the assumption that the propeller does not induce any tangential velocity, u, ahead of itself. Under the assumption of continuity of velocity across the propeller disk the Bernoulli equation for planes at the propeller and far downstream is

$$\frac{\rho}{2}\left[V^2(1-a)^2 + w^2 + u^2\right] + p + p' =$$

$$= \frac{\rho}{2}\left[(V-v_1)^2 + w_1^2 + u_1^2\right] + p_1 \qquad (2)$$

When the equations are solved for p' using the assumptions of $u = u_1$, $p_0 = p_1$, and $w = w_1 = 0$ it is seen that

$$p' = -\rho v_1 (V - v_1/2) \tag{3}$$

which from overall momentum flux considerations leads to the result

$$v_1 = 2aV \tag{4}$$

The thrust on an annular area $2\pi r dr$ is

$$dT = 4\pi r \rho V^2 a(1-a) dr \tag{5}$$

The moment of momentum $2r^2 a'\Omega$ imparted to the fluid equals the torque so that the power is given by

$$dP = 4\pi r^3 \rho V \Omega^2 a'(1-a) dr \tag{6}$$

From the relations shown in Figure 1 it is seen that for N blades of chord c

$$dT = \frac{\rho}{2} W^2 Nc(C_L \cos\phi + C_D \sin\phi)\, dr \tag{7}$$

and

$$dP = \frac{\rho}{2} W^2 Nc(C_L \sin\phi - C_D \cos\phi)\, r\,\Omega\, dr \tag{8}$$

When the corresponding relations are equated

$$\frac{a}{1-a} = \lambda(1 + \epsilon\tan\phi)\cot\phi\csc\phi \tag{9}$$

and

$$\frac{a'}{1+a'} = \lambda(1 - \epsilon\cot\phi)\sec\phi \tag{10}$$

with

$$\lambda = \frac{NcC_L}{8\pi r} \tag{11}$$

being the blade loading coefficient. The local power coefficient is given by

$$c_p = \frac{dP/dr}{2\,\pi r \rho V^3/2} = 4a(1-a)X\,\frac{\tan\phi - \epsilon}{1 + \epsilon\tan\phi} \tag{12}$$

with $X = r\Omega/V$, the tangential speed ratio, and $\epsilon = C_D/C_L$ the drag to lift ratio.

Optimum design, or maximum power output, is achieved when the right hand side of the equation is maximized for a given value of X. With the geometrical relation

$$\cot\phi = X\,\frac{1 + a'}{1 - a} \tag{13}$$

and elimination of a' from Equations 9 and 10 yields

$$X\tan\phi = 1 - \frac{a\sec^2\phi}{1 + \epsilon\tan\phi} \tag{14}$$

which is useful in the derivation of the relation for a corresponding to maximum c_p. After straightforward but tedious algebraic operations the maximum c_p is found to correspond to

$$a^{-1} = 2 + \frac{H}{2} + \left[\tan^2\phi + \left(1 + \frac{H}{2}\right)^2\right]^{\frac{1}{2}} \tag{15}$$

with

$$H = \epsilon\,\frac{\sec^2\phi}{T_1} + \frac{\epsilon\tan\phi\sec^2\phi}{T_1}$$

$$- \frac{2a - 1}{a(1-a)}\tan\phi\left(\epsilon T_2 - \epsilon^2 T_3\right)$$

and

$$T_1 = \tan\phi - \epsilon$$
$$T_2 = 2(\sin^2\phi + X\sin\phi\cos\phi)$$
$$T_3 = \tan\phi(\cos^2\phi - \sin^2\phi - 2X\sin\phi\cos\phi)$$

Considerable simplification is achieved with the approximation $1 + \epsilon\tan\phi \approx 1$ because only the first two terms appear in H. This then corresponds to the problem considered by Stewart (3). Another useful relation is obtained through the elimination of a from Equation 14 which leads to

$$\tan\phi = \frac{\sin\phi - (X^* - \epsilon)\lambda^*}{(X^* - \epsilon)\sin\phi + \lambda^*} \tag{16}$$

with

$$X^* = X(1 + \epsilon\tan\phi); \quad \lambda^* = \lambda(1 + \epsilon\tan\phi)$$

In terms of the solution vector $S = (\phi, \lambda, a)$ Equations 16, 15 and 9 may be viewed as an operator Φ on S, i.e.,

$$S = \Phi(S) \tag{17}$$

Convergence of the iteration scheme

$$S_i = \Phi\left(S_{i-1}\right) \tag{18}$$

implies a contraction mapping which by the Hahn-Banach fixed point theorem (e.g., Saaty, Nonlinear Mathematics, 4) assures the existence of a unique solution. These relations were programmed on a small computer and comprehensive parametric studies were performed over wide ranges of the drag/lift ratios ϵ and X.

PERFORMANCE OF BLADE ELEMENTS

Optimum performance of blade elements with finite drag/lift ratios is of great importance in preliminary performance estimates because it represents a standard against which actual blades must be measured. The measure of greatest significance is the power coefficient c_p which is shown in Figure 2 as a function of the tangential speed ratio X for a range of values of ϵ. It should be noted that peak values of c_p tend to occur at lower values of X as ϵ increases. However, even very efficient blade sections with $\epsilon = 0.01$ exhibit a maximum at $X \approx 3$, in contrast with ideal blades of $\epsilon = 0$ whose maximum c_p tends to 16/27 as X increases without bound. Similar trends are seen in the variations of the axial interference factor a with X which are shown in Figure 3. The torque grading coefficient which is defined as

$$c_T = c_p X^{-1} \tag{19}$$

is shown in Figure 4. In contrast with the very large influence of ϵ shown in the Figures cited above, Figures 5 and 6 indicate that the blade loading coefficient λ and the relative flow angle ϕ are rather insensitive to large variations in ϵ, particularly at lower values of X. These results are of considerable interest because they form the basis for the approximate analytical relations which are presented below.

OPTIMUM PERFORMANCE OF PROPELLER TURBINES

Wind turbine blades designed for optimum performance require a variation of the airfoil sections along the radius in order to ensure that each element is operating at its minimum drag/lift ratio. Thus, for any particular value of

ϵ, performance of a complete blade corresponds to an interval of one of the curves in Figure 2. This interval starts at a very low value of X corresponding to the hub and terminates at $X = X_0$ the tip speed ratio of the wind turbine which is customarily used to characterize particular designs. Actually, near the hub structural considerations dominate the problem and quite often no attempt is made to gain any aerodynamic performance in that region. It is quite common to use different airfoil sections in the inner and outer blade regions, but such fine points of optimization will not be considered here. The effects of tip losses and hub blockage will also be neglected since they are strong functions of particular designs and generally represent losses of only few percent. If the overall power coefficient C_P is defined as the power P normalized by the energy flow rate through a disk of radius R then

$$C_P = \frac{P}{\pi R^2 \rho V^3/2} = \frac{2}{X_0^2} \int_0^{X_0} X c_p \, dX \tag{20}$$

with $X_0 = R\Omega/V$ the integrand is already known from the previous section so that only simple quadratures are involved in the evaluation of C_P. Figure 7 shows the influence of the tip speed ratio, X_0, on the turbine power coefficient. The peaks of the curves are optimum tip speed ratios for turbines whose blades have a pitch distribution along the radius which corresponds to the optimum operation for that particular tangential speed ratio, X, and ϵ. It should be noted that the maxima of the overall power coefficients in Figure 7 occur at significantly higher values than those in Figure 2. It should also be noted that the variation of C_P with X_0 increases strongly with increasing ϵ which indicates high sensitivity of inefficient blades to off-design conditions. The distribution of power with the radius is shown in Figure 8. It should be noted that only about 5% of the total power is produced in the disk area corresponding to 20-25% of the radius.

The distribution of λ with the dimensionless radius is shown in Figure 9. For representative airfoil sections minimum ϵ is achieved at values of C_L of .5 - .8 (e.g., Abbott and von Doenhoff, Theory of Wind Sections, Reference 5) so that the solidity $Nc/2\pi r$ is approximated by $\sim 7\lambda$. For turbines with $\epsilon = 0$ and $X_0 = 10$ this corresponds to solidity of about 1.5% near the tip. Inefficient blade sections with $\epsilon = .1$ and $X_0 = 2$ would have corresponding solidities of 45%. Obviously, the results which indicate solidities approaching 100% must be disregarded since they are in error due to the limitations of the theory. It must be stressed that the purely aerodynamic considerations of blade area distributions are inevitably modified by structural considerations which introduce centrifugal loads, aeroelastic effects and fatigue life.

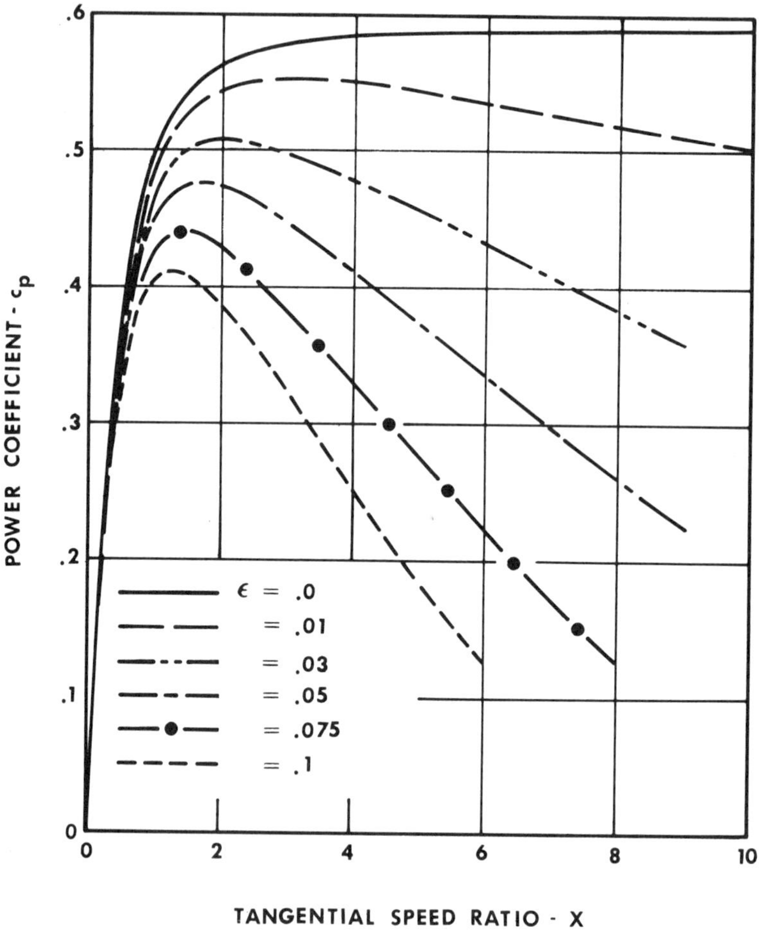

Figure 2. Variation of the power coefficient with tangential speed ratio

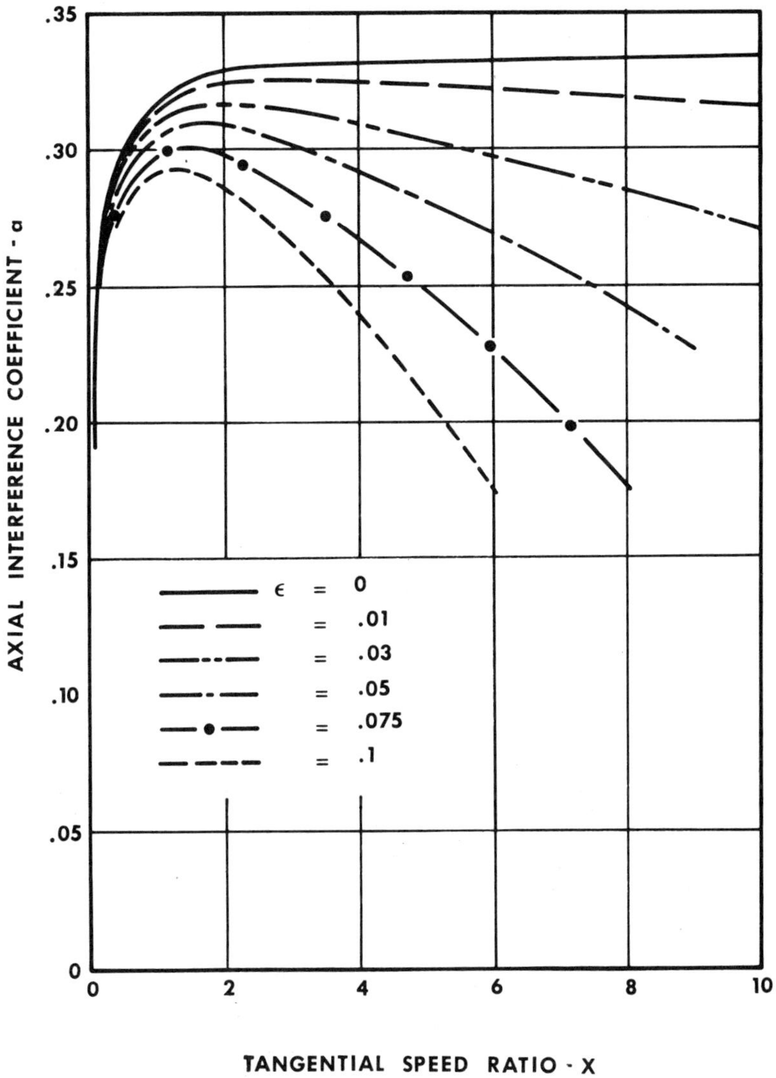

Figure 3. Variation of the axial interference factors with the tangential speed ratio

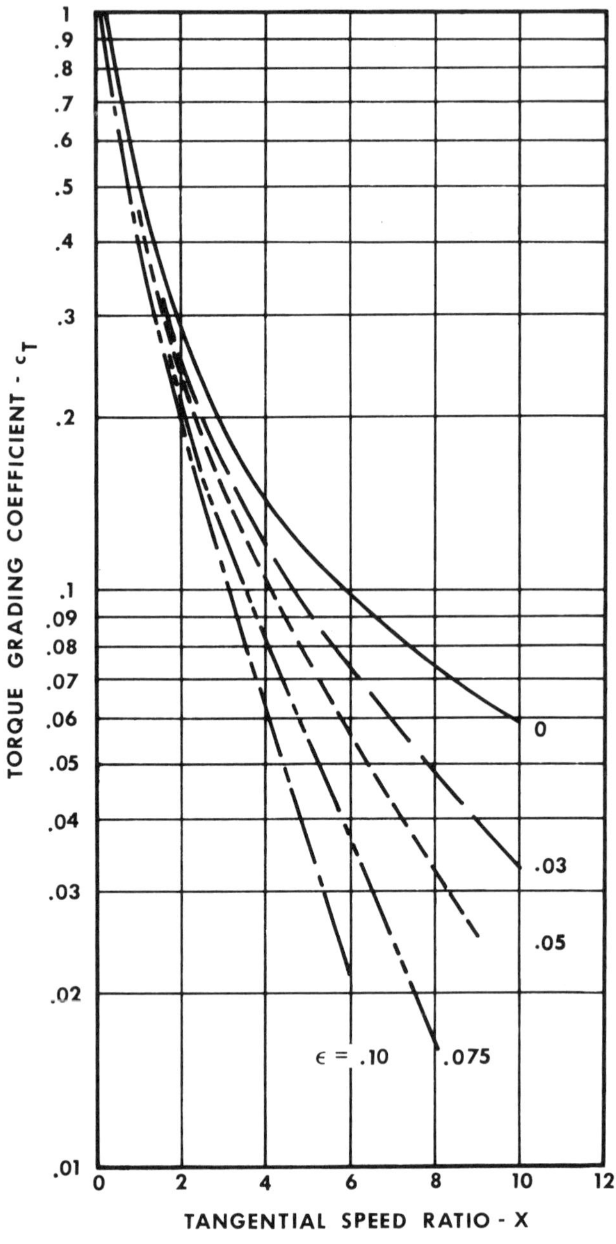

Figure 4. Variation of the torque grading coefficient with the tangential speed ratio

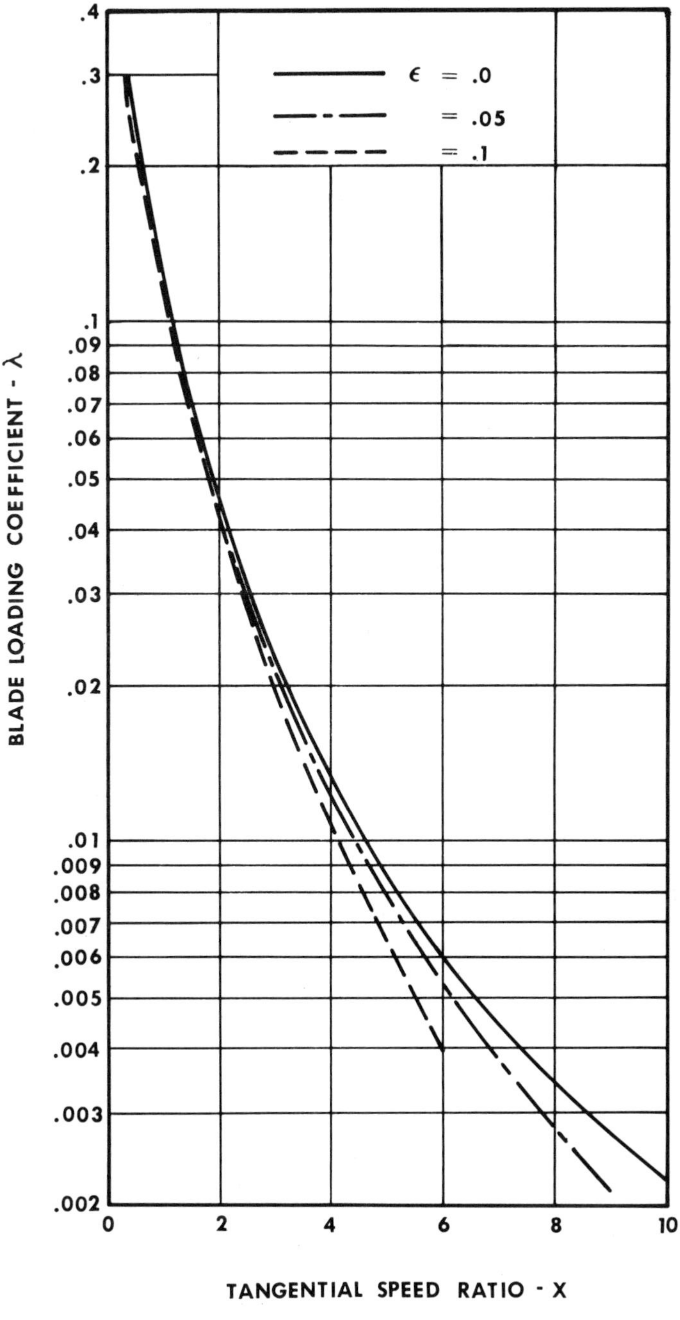

Figure 5. Variation of the blade loading coefficient with the tangential speed ratio

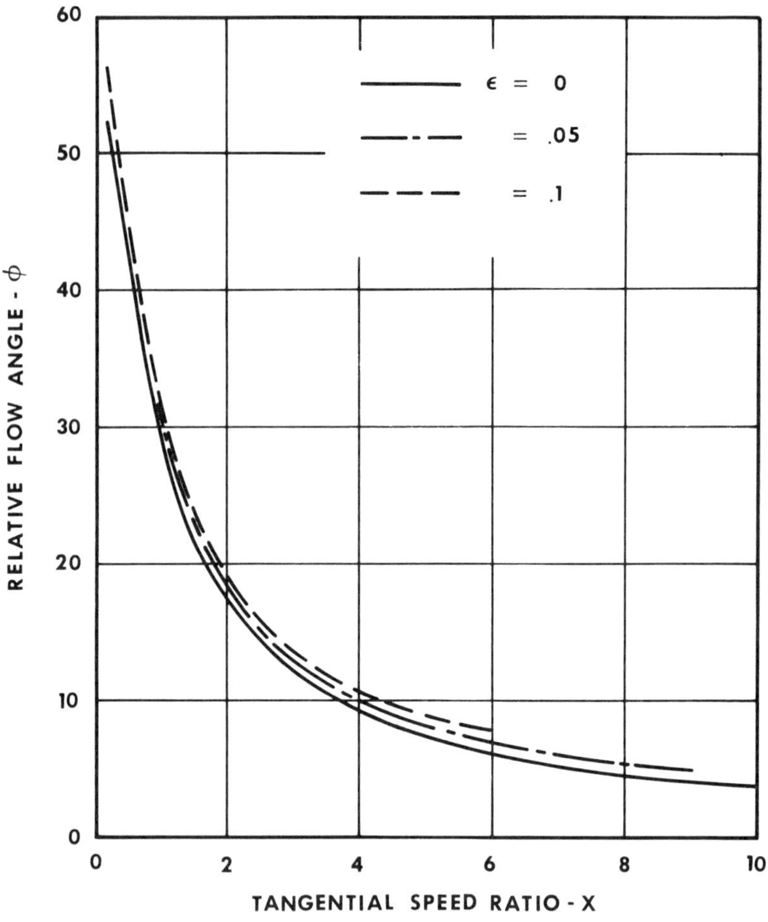

Figure 6. Variation of the relative flow angle with the tangential speed ratio

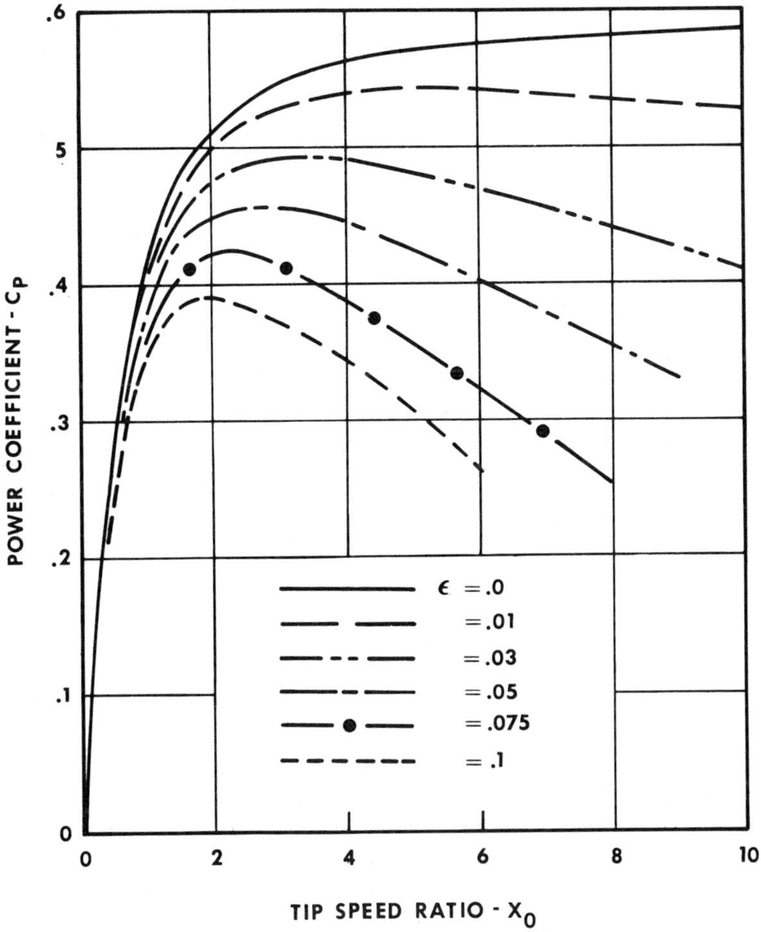

Figure 7. Influence of the tip speed ratio on the turbine power coefficient

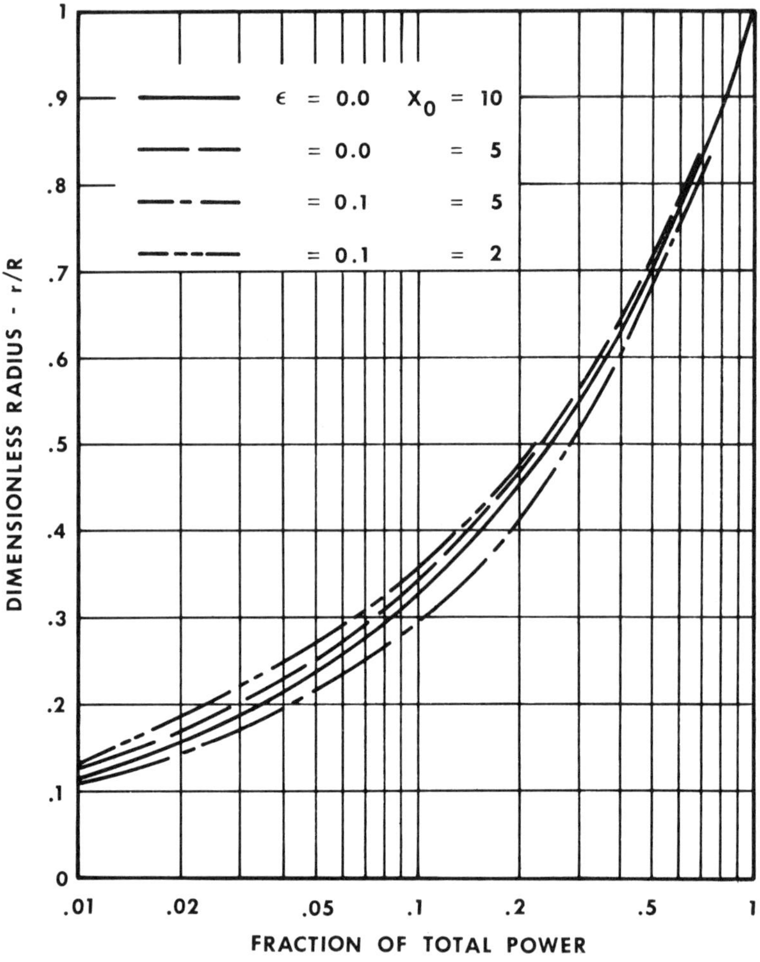

Figure 8. Distribution of power
extracted with radius

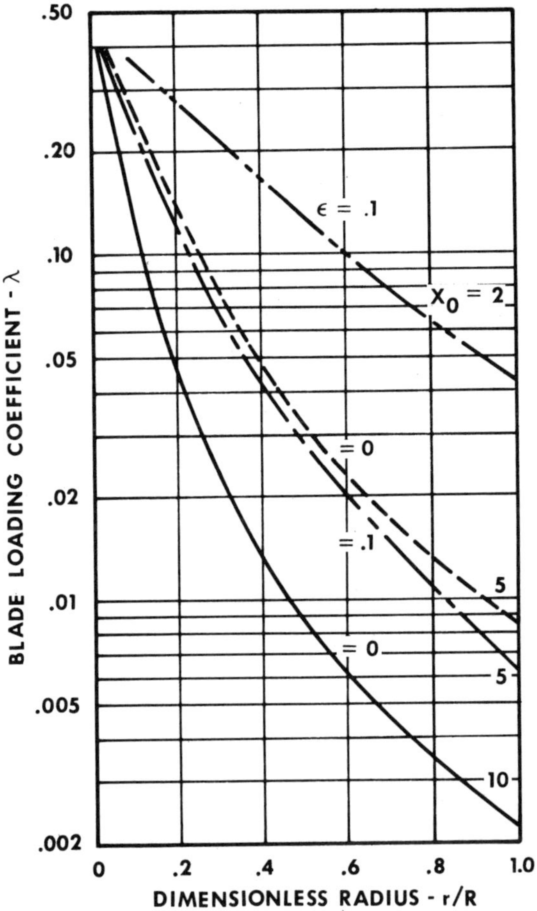

Figure 9. Distribution of the blade loading coefficient with radius

FORCES AND MOMENTS ON BLADES

One of the primary considerations in the design of wind turbines is the assurance of the structural integrity of the components during the design life. It is therefore necessary to know the forces and moments on blades. The axial force on the annular elements in all the blades is seen from Figure 1 to be

$$df_a = 4\pi r a(1-a)\rho V2dr \tag{21}$$

The distribution of the force coefficient C_{Fx} is given by

$$C_{Fx} = \int_z^1 \frac{df_a}{\pi R^2 \rho V^2/2} = 8\int_z^1 za(1-a)\,dz \tag{22}$$

with $z = r/R$

Representative examples of calculated values of C_{Fx} are shown in Figure 10. It should be noted that the very efficient blades tend towards the limiting value of 8/9.

The principal moment on the blades is due to the axial force. This is calculated here as a moment coefficient C_{Mx} referenced to the turbine parameters, as follows

$$C_{Mx} = \int_z^1 \frac{zdf_a}{\pi R^3 \rho V^2/2} = 8\int_z^1 z^2 a(1-a)\,dz \tag{23}$$

Representative examples of the distribution of C_{Mx} with z are shown in Figure 11. Again, the very efficient blades tend to the theoretical limit of 16/27.

The axial force and the moment due to it which are considered here enter into stress calculations of blades together with the torque and centrifugal forces. When all these are taken into account the optimum blade design may not necessarily be aerodynamically optimum.. Examples of that may be seen in some airplane propellers which employ very thick elliptical sections near the hub to withstand centrifugal stresses and torque. Distribution of aerodynamic loading along the blade is subject to similar concerns and compromises between aerodynamic performance and structural integrity are necessary.

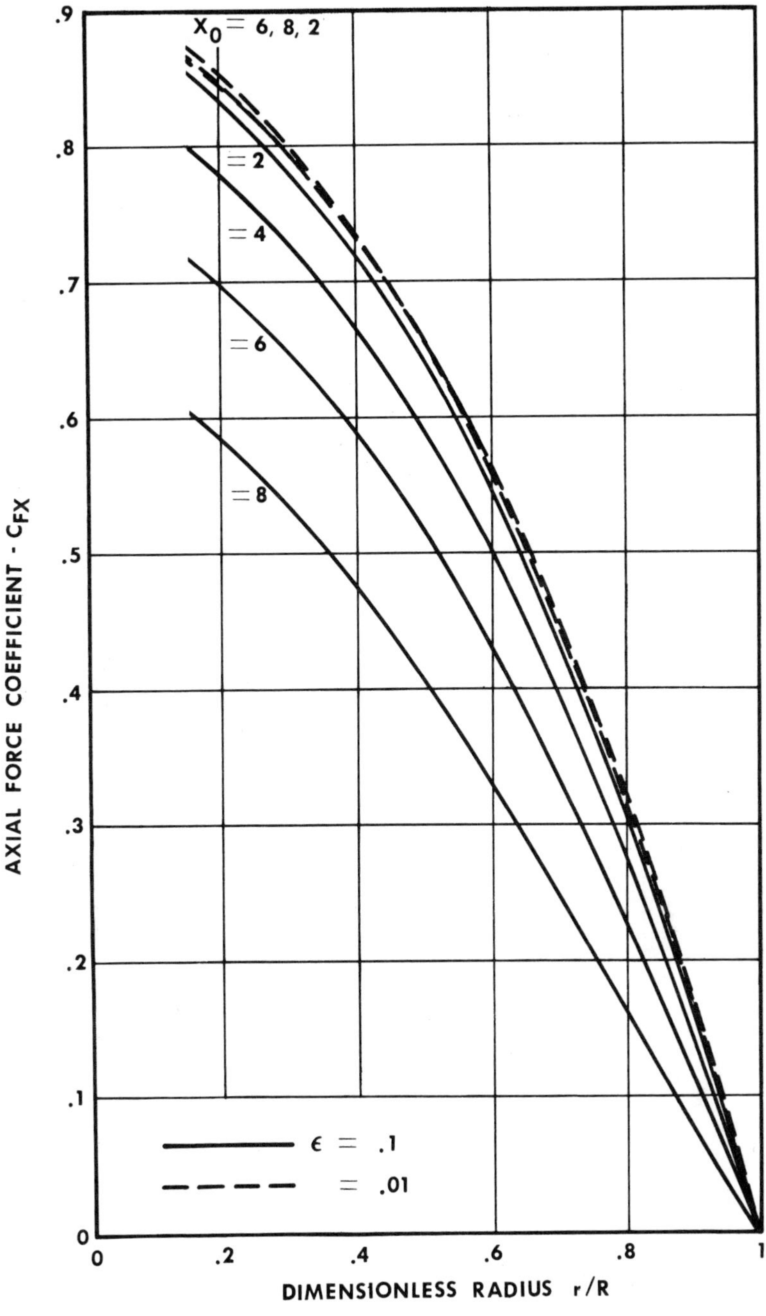

Figure 10. Distribution of axial force coefficients under optimum operating conditions

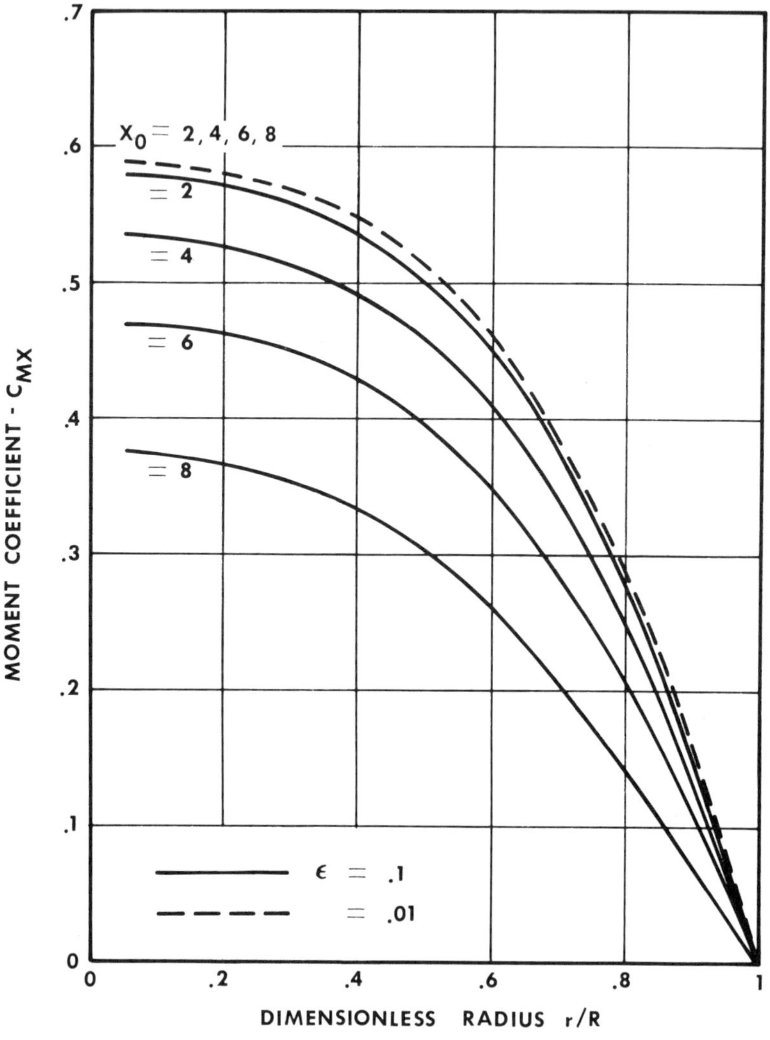

Figure 11. Distribution of the moment coefficients resulting from axial forces with dimensionless radius

AERODYNAMIC PERFORMANCE OF BLADES

The optimum performance results presented in the preceding sections were all based on the assumptions that the drag/lift ratio ϵ is a given parameter of the problem. However, because of the variability of winds and the frequent demand for a constant angular rate operation, wind turbines rarely operate at their design conditions. Even with variable pitch blades off-design operation will occur when wind speeds vary significantly. It is therefore necessary to know the variation of the aerodynamic characteristics of the blade elements with angle of attack and to a certain degree the variation with the Reynolds number when the combination of relative wind speeds and blade chords results in operation in the range of Reynolds numbers in which significant variations are encountered. In airplane wings the variation of drag/lift ratios and lift coefficients with angle of attack can be estimated using semi-empirical formulas when the aspect ratio and the zero-lift drag coefficient are known. In propeller wind turbines the problem is much more difficult because it is not possible to assign an effective aspect ratio to a rotating blade. However, according to von Mises (2) experience indicates that the effective aspect ratio for airplane propellers is between 10 and 20. In the following the effective aspect ratio A_R and the minimum drag/lift ratio ϵ_0 will be used as convenient parameters for systematic characterization of blade element aerodynamic performance. For detailed calculations, particularly if stall angles were to be approached, it would be necessary to use actual aerodynamic performance data.

The slope of the lift curve of a wing with aspect ratio A_R = Span^2/Area is given by Hoerner (6) as

$$C_{L_\alpha} = 2\pi\ddot{a}/(1 + 2\ddot{a}/A_R) \tag{24}$$

The induced drag is given by

$$dC_D / dC_L^2 = 1/\pi A_R \tag{25}$$

so that

$$C_D = C_L^2 / \pi A_R + C_{D_0} \tag{26}$$

with C_{D_0} being the drag coefficient at zero lift. The drag/lift ratio is therefore

$$\epsilon = C_D / C_L = C_L / \pi A_R + C_{D_0} / C_L \tag{27}$$

It is convenient to characterize the performance of wings in terms of the minimum drag:lift ratios, and aspect ratios. In those terms, the minimum drag is given by

$$C_{D_0} = \pi\epsilon_0^2 \mathcal{R} /4 \tag{28}$$

and the angle of attack for minimum ϵ

$$\alpha_0 = C_{L,0} / C_{L_\alpha} \tag{29}$$

with the lift coefficient at minimum ϵ, $C_{L,0}$, being given by

$$C_{L,0}^2 = \pi C_{D_0} \mathcal{R} \tag{30}$$

The variation of ϵ with C_L may now be expressed as

$$\frac{\epsilon}{\epsilon_0} = \frac{1}{2}\left[\left(C_L / C_{L,0}\right) + \left(C_L / C_{L,0}\right)^{-1}\right] \tag{31}$$

In the following it will be assumed that the blade airfoils will be operating well below the stall point so that linear variation of C_L with α will be acceptable. Representative examples of calculated values of ϵ together with some data from Abbott and von Doenhoff (5) are shown in Figure 12.

It must be emphasized that the use of the aspect ratio as a parameter in the characterization of airfoil performance in propeller blade analysis is merely an artifice which gives an adequate representation of the magnitudes and variations with angle of attack of the drag to lift ratio parameter. In any detailed preliminary design studies, it would be necessary to employ experimentally determined airfoil performance data with Reynolds number effects. Even then, the corrections for radial flow on a rotating blade introduce some uncertainty into the results. This would be particularly true in cases when local blade sections are approaching stall conditions. The recommended design procedure is to use high aspect ratio wing data together with blade pitch angle distribution which will assure operation well below the stall points of the sections.

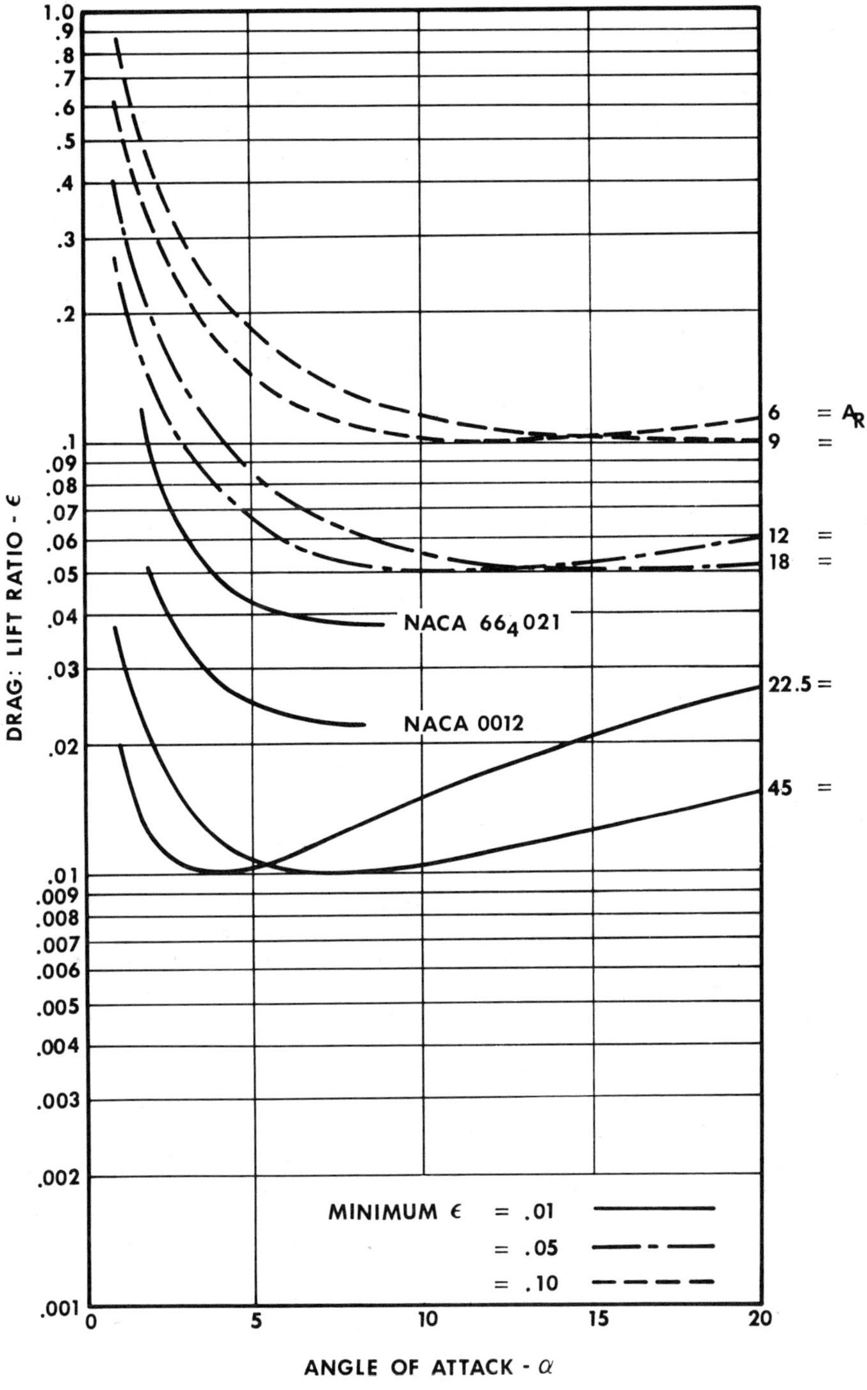

Figure 12. Performance of representative sections

OFF-DESIGN PERFORMANCE

Operation of a propeller wind turbine at its design point means that all the blade sections will be pitched in such a way that minimum drag/lift ratio is realized. In situations when the angular rate of the turbine may be varied and the demands of the load are compatible it is possible, in principle, to achieve optimum operation in a range of wind speeds. The variations in Reynolds numbers in such instances introduce deviations from the absolutely optimum performance, but these are of secondary importance in larger wind turbines. When wind turbines are used to generate electricity, it is usually advantageous to deliver power at a fixed voltage and frequency in order to permit the use of standard electrical devices. Larger machines are linked to the main electrical power grid so that frequency must be synchronized precisely. Large propeller wind turbines achieve such control by variation of the pitch of the blades. Smaller wind turbines which must operate at a fixed angular rate operate at off-design conditions whenever wind speeds deviate from the nominal value. The situation may be alleviated somewhat by the recent developments of electronically controlled variable input speed, constant frequency output generators. At present these are limited in output to 50-60 kW and are very expensive because they are intended for use in aircraft. With the usual development of such devices, the costs are bound to decrease, particularly if non-aircraft applications are found which will permit some relaxations of standards for weight reliability and life of the equipment.

This section illustrates the variation of the power coefficients of fixed pitch turbines with the variations of the tip speed ratio. Relative changes in power coefficients with relative changes in speed ratios are shown in Figure 13 for a range of designs. It should be noted that a speed ratio greater than unity means that a constant angular rate turbine is operating in wind speeds below the nominal value. The Figure indicates that considerable latitude exists in the combinations of airfoil section characteristics and design tip speed ratios which will result in machines with high operating efficiencies in a wide range of wind speeds.

The section on optimum performance showed that the maximum power coefficient for a rotor was achieved at increasing values of the tip speed ratios as the drag/lift ratio diminished. These maxima tend to be less sharply defined at lower values of ϵ, but are readily discernible. Designs of wind turbines employing some selected airfoil sections with a particular ϵ over the whole length of the blade would select tip speed ratios corresponding to the maxima in the curves of Figure 7. Operation of such turbines in off-design conditions is an interesting engineering problem which influences very heavily the overall system consideration. For the purposes of illustration, the examples selected here are for airfoil sections which achieve their minimum drag/lift ratios, ϵ_0, at angles of attack α_0, of 0.1 radians, or 5.73°. The three values of ϵ = 0.01, 0.03, 0.05 are associated with design tip speed ratios of 5, 3.5 and 3 respectively which correspond approximately to the maxima in the curves in Figure 7. Distribution of the pitch angles with the radius is shown in Figure 14. The variation of the power coefficient with the tip speed ratio which is shown in Figure 15 illustrates the extreme sensitivity of relatively inefficient turbines to variations in operating conditions. Such curves are strong functions of the airfoil sections so that the results shown here serve to illustrate trends, and indicate the procedures which must be followed in order to arrive at an acceptable design.

AUGMENTED WIND TURBINES

The low energy density of winds and the relatively low theoretical upper limit on the extraction of power have stimulated considerable interest in augmentation of power output of rotors. The numerous augmenting schemes which have been proposed may be grouped into three major classes:

1. Concentrators
2. Aerodynamic lift auxiliaries
3. Diffusers - ejectors

Within each group there seems to be a limitless number of variations as new ideas appear continuously. Typical concentrators channel large quantities of flow into the rotor and thus increase its power output because both the flow rate and the flow speed are increased. At first glance this appears to be an easy method for increasing the power output of a rotor but upon closer examination, it is seen that the overall momentum and energy balances limit the power coefficient referenced to the projected inlet area to the Betz-Lanchester limit of 16/27. In fact, because of the inlet and frictional losses such a limit cannot be reached even theoretically. The reason for this is due to the fact that the flow exiting the rotor must be diffused to ambient pressure, and this sets the upper limit on the flow rates which can be handled. Concentrators must be viewed as attempts to gain additional power from a given rotor through the use of passive, and presumably inexpensive, aerodynamic surfaces. The disadvantages are that the additional weight of the system would add so much to the tower and foundation cost that a ground plane location might have to be considered. In such a case, the system would incur the penalty of operating in lower wind speeds, and probably not being able to point into the wind. There does not appear to be any reasonable justification for such an approach.

Auxiliary devices which employ aerodynamic lift to augment the power output of rotors exhibit an even greater range of imagination than can be found in the field of concentrators of the ideas being promoted are aerodynamically unsound, but some concepts merit some attention. Considerable attention has been focused on the use of vortices to create regions of low pressures and high velocities. These vortices can be generated by buildings, wind tips or delta wings. Loth (7) examined several variants of this approach and concluded that only limited benefits could be realized in some cases. A more promising approach seems to be in the direction of using tip vanes which are aerodynamic surfaces placed on the tips of the blades at such an angle that their lift tends to turn the wind into the rotor. No conclusive demonstration of the effectiveness of tip vanes has appeared, but preliminary studies by van Holten (8) indicate the existence of a considerable potential for augmenting the power output of a rotor.

The approach of using ejectors or ejector-diffusers, depending on the particular configuration adopted attempts to augment the power output by increasing the pressure drop across the rotor. The potential effectiveness of such augmentors is readily estimated when the Bernoulli equation is written for the flow ahead of the rotor and then for the downstream flow in the same form as was used before

$$p_0 + \rho v_0^2/2 = p' + \rho v'^2/2 \tag{32a}$$

$$p_1 + \rho(v_0 - q)^2/2 = p'' + \rho v'^2/2 \tag{32b}$$

with p_0, v_0 and p_1, v, being the pressures and velocities very far ahead and very far behind the rotor respectively. The velocity deficit at the location of p_1 is denoted by q. Velocity at the rotor is denoted by v' and p' and p" are pressures immediately ahead and immediately behind the rotor respectively. When the equations are solved for the pressure drop without the assumption of $p_0 = p_1$ then

$$p' - p'' = \frac{\rho v_0^2}{2}\left[\frac{2q}{v_0}\left(1 - \frac{q}{2v_0}\right) - c_{pr}\right] \tag{33}$$

with the exit pressure coefficient c_{pr} defined as

$$c_{pr} = (p_1 - p_0)/(\rho v_0^2/2)$$

Under the previously discussed conditions of negligible thrust by ambient pressure on the streamtube going through the rotor the momentum equation for a unit area of the rotor is written as

$$p_0 + \rho v' v_0 = T + p_1 + \rho v'(v_0 - q) \qquad (34)$$

with T being the thrust per unit area which when equated to the pressure drop leads to the familar result

$$\frac{v'}{v_0} = 1 - \frac{1}{2}\frac{q}{v_0} \qquad (35)$$

The power is the rate of work done by the fluid moving through the rotor, Tv', and the power coefficient referenced to the free stream speed and rotor area becomes

$$C_P = 4a(1 - a)^2 - (1 - a)c_{pr} \qquad (36)$$

For a given exit pressure coefficient, c_{pr}, the maximum power coefficient is found to occur at a value of the interference factor, a, given by

$$a = \frac{2}{3} - \frac{1}{3}\left(1 - \frac{3}{4}c_{pr}\right)^{\frac{1}{2}} \qquad (37)$$

which under the usual conditions of ambient pressures at initial and final planes reduces to the Betz-Lanchester limiting value of 1/3. Optimum performance of wind turbines with reduced exit pressures is shown in Figure 16. It should be noted that the interference factor is reduced with decreasing exit pressure coefficient and can, in fact, become negative which means that the flow into the turbine would be accelerated.

The apparently high potential gains from the use of diffuser-ejector schemes turn out to illusory when a complete system study is performed. In Reference 9 a detailed study was made of a wind turbine within a shroud which acted as an ejector with a constant area mixing duct followed by a diffuser. In essence, a completely general augmentation system for wind turbines was analyzed in terms of energy and momentum balances. A schematic representation of the system is shown in Figure 17. The results of the study indicated that while the power output of the turbine could be augmented by factors of 10-20, the overall power coefficient referenced to the enormous diffuser area was inferior to that of a free turbine. The enormous sizes and weights of such devices confined them to ground level installation which reduced the available power because of lower wind speeds. A 1 megawatt device was calculated to require a diffuser exit diameter of 230-310 ft and an overall length of about 1400 ft. It would be quite difficult to point such large machines into the wind so that the energy capture would be inferior to that of a free turbine which can be yawed quite readily.

In conclusion, it appears that the concentrator and diffuser-ejector concepts are basically impractical. The best vortex schemes appear to be marginal. The one approach which seems to offer some encouragement appears to be in the area of tip vanes. The benefits are likely to be quite modest, but the additional cost and complications are relatively small.

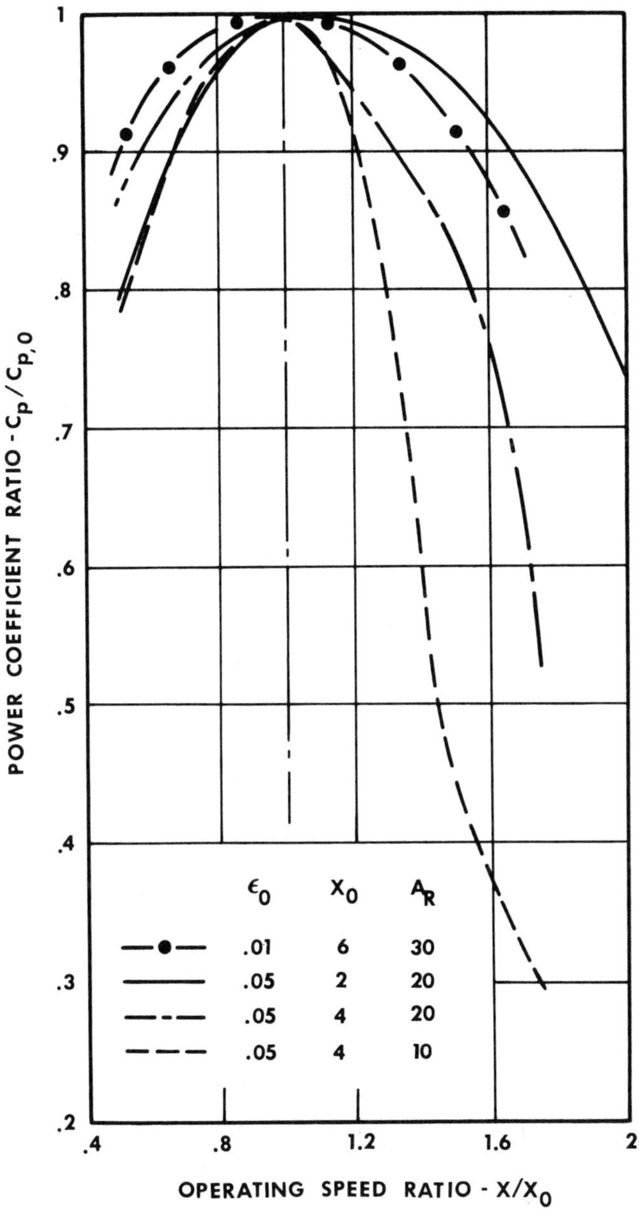

Figure 13. Relative changes of power coefficients with relative tip speed changes

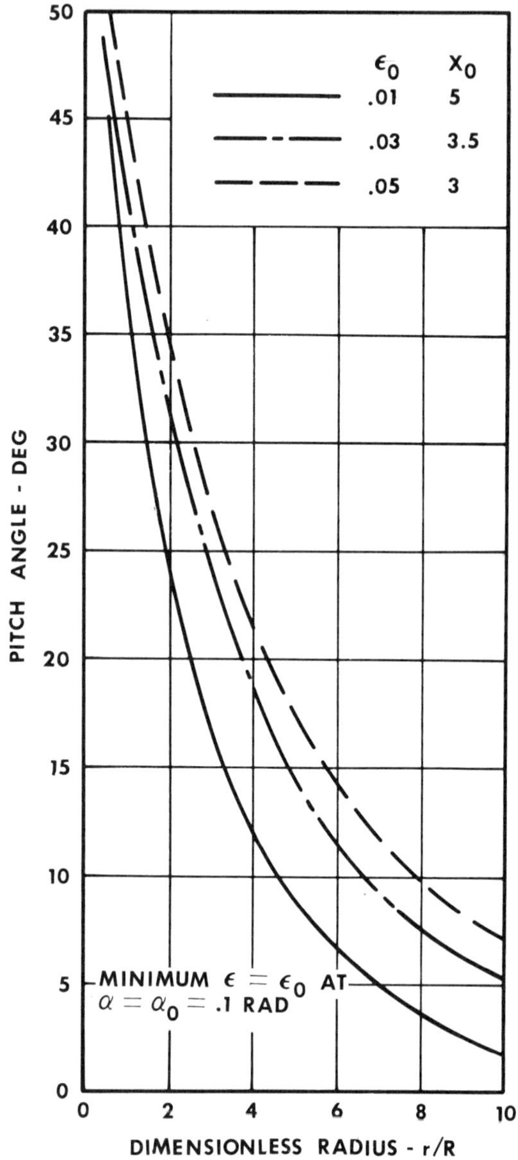

Figure 14. Variation of pitch angles with radius

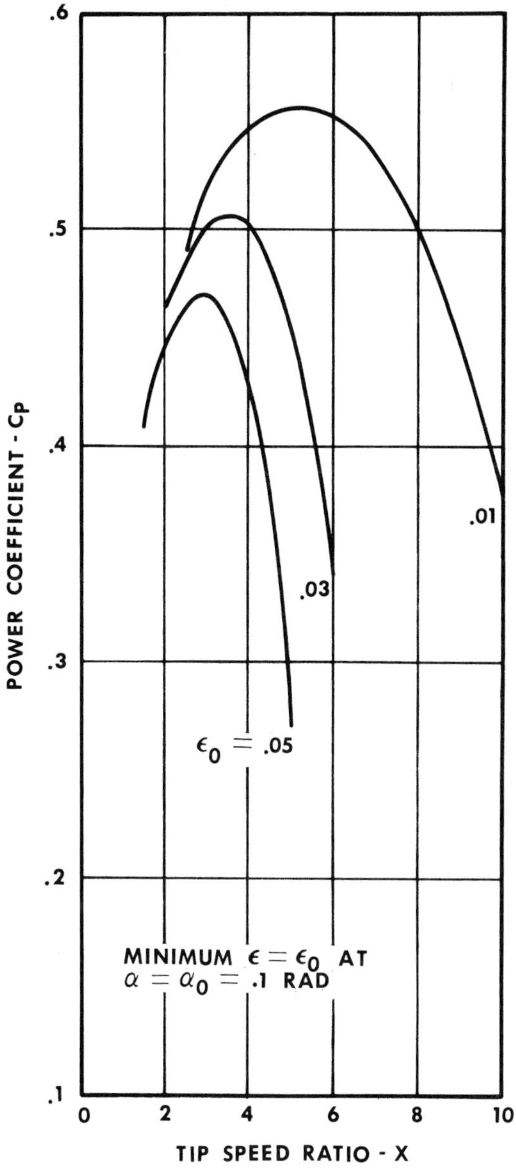

Figure 15. Variation of the power coefficients of rotors with tip speed ratio

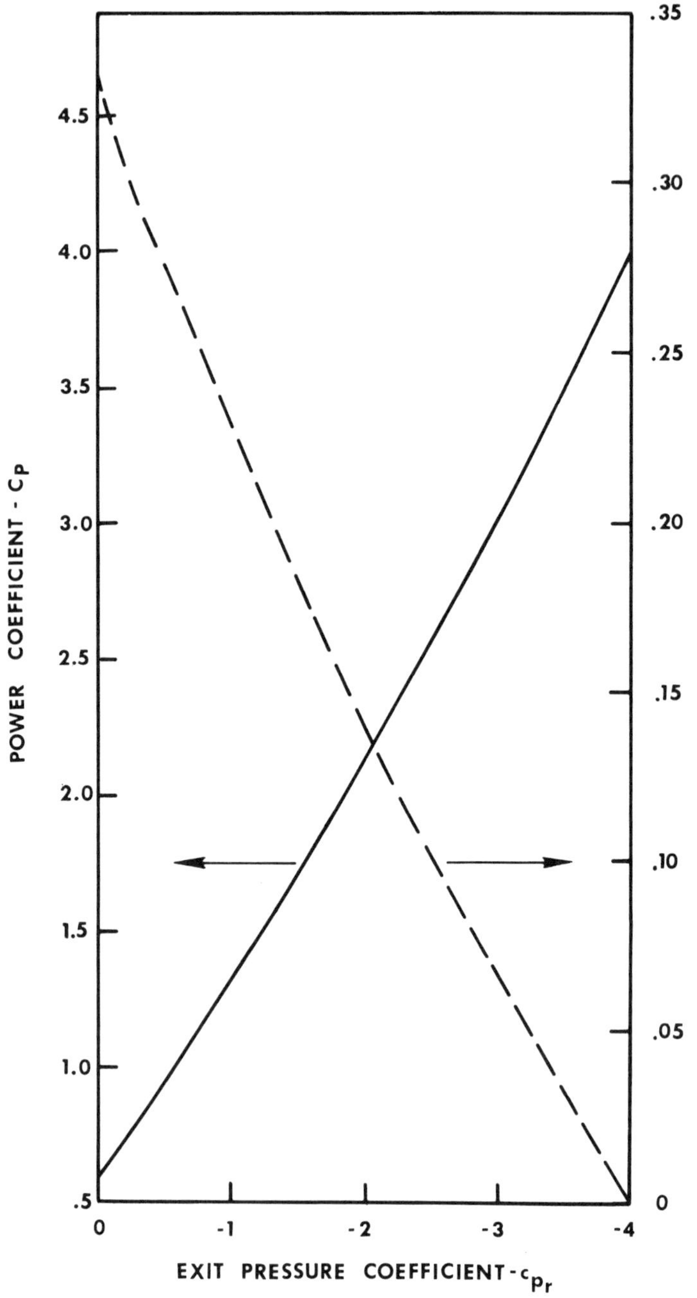

Figure 16. Augmentation of the power coefficient by reduction of the exit pressure

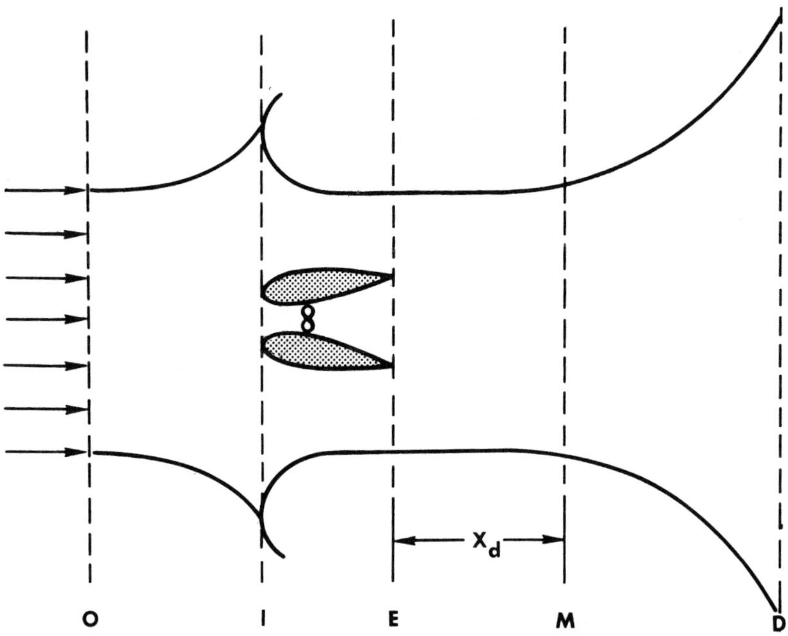

Figure 17. Schematic representation of shrouded ejector turbine

APPROXIMATE RELATIONS

Approximate relations are often quite useful for quick estimates, even if computer codes, such as the ones used to calculate results here, are available. In order to be useful an approximation must be so simple that functional relationships are readily perceived and calculations can be performed with a minimum of effort. The basis for the approximations will be Equations 14 and 15 which in the more usual reduced form are written as

$$X\tan\phi = 1 - a\sec^2\phi \tag{38}$$

$$a^{-1} = 2 + \frac{H'}{2} + \left(\frac{\tan\phi\sec^2\phi}{\tan\phi - \epsilon} + H'\right)^{\frac{1}{2}} \tag{39}$$

with

$$H' = \frac{\epsilon\sec^2\phi}{2(\tan\phi - \epsilon)}$$

Elimination of the cosine terms from Equation 38 by substitution of trigonometric identities leads to the result

$$\sin^2\phi = \frac{1}{4}\left[1 - \left(1 - \frac{4(1-a)^2}{X^2}\right)^{\frac{1}{2}}\right] \tag{40}$$

which may be simplified to an approximate form

$$\sin^2\phi \approx \frac{(1-a)^2/X^2}{1 - (1-a)^2/X^2} \tag{41}$$

and simplified further to

$$\sin\phi = \frac{1-a}{X}\left[1 + \frac{1}{2}\frac{(1-a)^2}{X^2}\right] \tag{42}$$

Approximation to the interference factor a is obtained by using the definition

$$a = (1 - \delta)/3$$

and solution of Equation 39 to first order for δ

$$\delta = \frac{1}{3}\left(\frac{2}{9X^2} + \frac{3\epsilon X}{2 - 3\epsilon X}\right) \tag{43}$$

At optimum operating conditions which are considered above

$$\lambda = 1 - \cos\phi \approx \frac{(1-a)^2}{2X^2} \tag{44}$$

It is now clear why ϕ and λ exhibit such weak influence of ϵ when the latter is relatively small. The power coefficient is obtained by straight-forward substitution for a and is written as

$$c_p = \frac{8}{27}\left(1 - \frac{\delta}{2}\right)\left(2 - 3\epsilon X + \delta\right) \tag{45}$$

which in the limit of vanishing ϵ and large X tends to the limiting value of 16/27. Through the use of the above approximate relations the axial force coefficient is found to be

$$C_{Fx} = \frac{8}{9}\left[1 - \frac{\epsilon X_0}{6} - \frac{3(\epsilon X_0)^2}{16}\right] \tag{46}$$

while its component from the tip to the dimensionless radius $z = r/R$ is

$$C'_{Fx} = \frac{8}{9}\left[(1 - z^2) - \frac{\epsilon X_0}{6}(1 - z^3) - \frac{3(\epsilon X_0)^2}{16}(1 - z^4)\right] \tag{47}$$

Similarly, the moment due to the axial force is calculated to be

$$C_{Mx} = \frac{16}{27}\left[1 - \frac{1}{9X_0^2} - \frac{3\epsilon X_0}{16} - \frac{9(\epsilon X_0)^2}{40}\right] \tag{48}$$

and its component form the tip to the dimensionless radius is

$$C_{Mx} = \frac{16}{27}\left[(1 - z^3) - \frac{1-z}{9X_0^2} - \frac{3\epsilon X_0}{16}(1 - z^4) - \frac{9(\epsilon X_0)^2}{40}(1 - z^5)\right] \tag{49}$$

With the above relations all other parameters of the problem can be calculated using the basic relations presented earlier in the Chapter. A quick check against the calculated data shown in the Figures will demonstrate that the approximate results presented here are quite acceptable for engineering estimates in preliminary analyses of projects.

CONCLUDING REMARKS

The subject of propeller wind turbines was covered in considerable detail here because this class of wind turbines has a fairly good theoretical basis in contrast to the other types of wind machines. Most of the results shown here are original and, at the present time, cannot be found elsewhere. Again, it must be noted that the theory on which the analyses are based is an approximation which has served well in the past and will have to do until complete numerical solutions for a finite number of blades and wake interference effects becomes generally available.

REFERENCES

1. Glauert, H., "Airplane Propellers," Aerodynamic Theory, W.F. Durand Ed., J. Springer, Berlin, 1935, pp. 324-332.

2. von Mises, R., Theory of Flight, Dover, New York, 1959, pp. 285-355.

3. Stewart, H. J., "Dual Optimum Aerodynamic Design for a Conventional Windmill," AIAA Journal, Vol. 14, No. 11, Nov. 1976, pp. 1524-1527.

4. Saaty, T. L., Modern Nonlinear Equations, McGraw-Hill, New York, 1967.

5. Abbott, I. H., and von Doenhoff, A. E., Theory of Wing Sections, Dover, N.Y., 1959.

6. Hoerner, S. F., and Borst, H. V., Fluid-Dynamic Lift, Hoerner Fluid Dynamics, 1975, p. 3-3.

7. Loth, J. L., "Wind Power Limitations Associated with Vortices," AIAA J. of Energy, Vol. 2, No. 4, July-August 1978, pp. 216-222.

8. van Holten, Th., "Concentrator Systems for Wind Energy, with Emphasis on Tip-Vanes", Wind Engineering, Vol. 5, No. 1, January 1981, pp. 29-45.

9. Wortman, A., "Evaluation of the Potential of Wind Turbines with Ejectors and Diffusers," Report AWD-12-80, Dec. 1980, AWD Inc., Santa Monica, CA.

CHAPTER 5

VERTICAL AXIS LIFTING SURFACE TURBINES

Vertical axis wind turbines overcome the difficulties of propeller windmills which must be pointed into the wind by orienting the axis of rotation normally to the wind so the full force of the wind can be sensed from any direction. In this Chapter considerations will be limited to vertical axis turbines which employ aerodynamic lifting surfaces to generate torque. The principal types of such turbines are the Darrieus rotors which have the troposkien or skipping rope curved blades and the straight bladed giromill types which may or may not vary the blade pitch during the cycle. Variants of the cross-wind axis types such as lifting surfaces on vehicles moving along a curved track or belt type devices have been proposed, but their superiority over current concepts has not yet been demonstrated. Advantages of vertical axis turbines are:

a. Elimination of yawing mechanisms
b. High efficiency
c. High rotational speeds which reduce gear ratios

In giromills with articulated blades it is necessary to sense the direction of the wind in order to time the cyclic pitch changes correctly. The disadvantages are:

a. Low starting torque
b. Tendencies to stall in gusts
c. Sensitivity to off-design conditions
d. Dynamic stability problems
e. Generally low height installation which results in operation in lower wind speeds

Rotation of the blades around the axis changes both the angle of attack of the aerodynamic surface and the local dynamic pressure. The torque produced by the blades is not steady so that the load driven by the turbine sees a fluctuating power input. In the case of Darrieus rotors and straight bladed machines without blade articulation mechanisms regulation of angular rate must be performed by the load.

Analysis of the aerodynamics of vertical axis turbines is extremely difficult because of the variability of the local flow conditions as the blades orbit around the axis. When viewed from some distance as an aerodynamic system a vertical axis wind turbine is a device which produces not only drag, but also lift in the direction normal to the wind vector. The resulting flow field is quite complex and it is very difficult to make accurate estimates of the inflow conditions into the turbine. In the windward part of the rotation cycle the blades move into a skewed flow field which varies rapidly with location. In one quadrant of the windward part of the cycle the blades are advancing into the wind while in the other they are retreating. In the leeward side of the cycle the blades are operating in a flow field influenced by the blade motion in the windward phase of the operation. The flow situation is thus extremely complex and at present beyond existing analytical capabilities. An attempt, based on numerous simplifying assumptions was made by Wilson, et al (1) who assumed only axial interference in the flow velocity and calculated power coefficients by following the analytical procedure from propeller theory of equating blade forces with wake momentum defect. Even though the analysis is based on a large number of assumptions, which in some instances contradict theoretical principles, the results indicate the trends correctly. Analyses employing similar, drastic assumptions have also been made by Templin (2), Muraca, et al (3), Sheldahl, et al (4), and Ashley (5). The latter has developed an approach which is probably as good as any of the other theories, but its ability to predict performance of Darrieus rotors is not entirely satisfactory. Figure 1 shows the comparisons between Ashley's calculations and experimental data. Also shown are the points generated by a curve fit for the form

$$C_P = C_{P,0} \exp\left\{-c\left[\left(\frac{X-X_0}{X_0}\right)^{\alpha} + \left(\frac{X-X_0}{X_0}\right)^3\right]\right\} \qquad (1)$$

with

$$C = 2.5$$
$$\alpha = 1.75$$

This curve will be used in the estimates of the annual output of such turbines in the following sections.

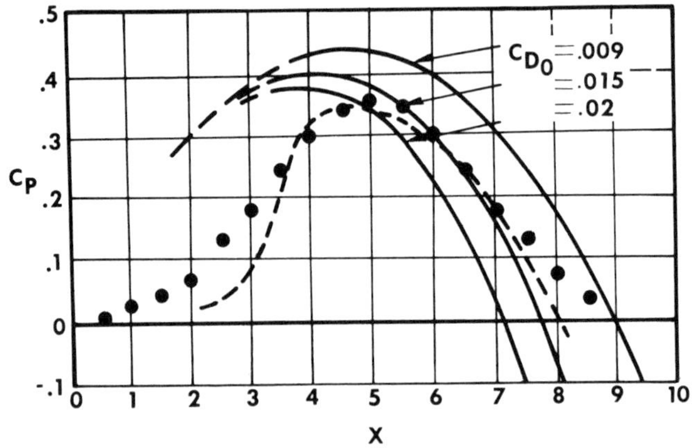

Figure 1. Performance of Darrieus rotors, comparison of Ashley's Theory ——, with experimental data – – –, and a curve fit ● ● ● ●

Some appreciation for the complexity of the flow phenomena involved in the operation of vertical axis turbines may be gained from the local flow vector diagram for a blade section which is shown in Figure 2.

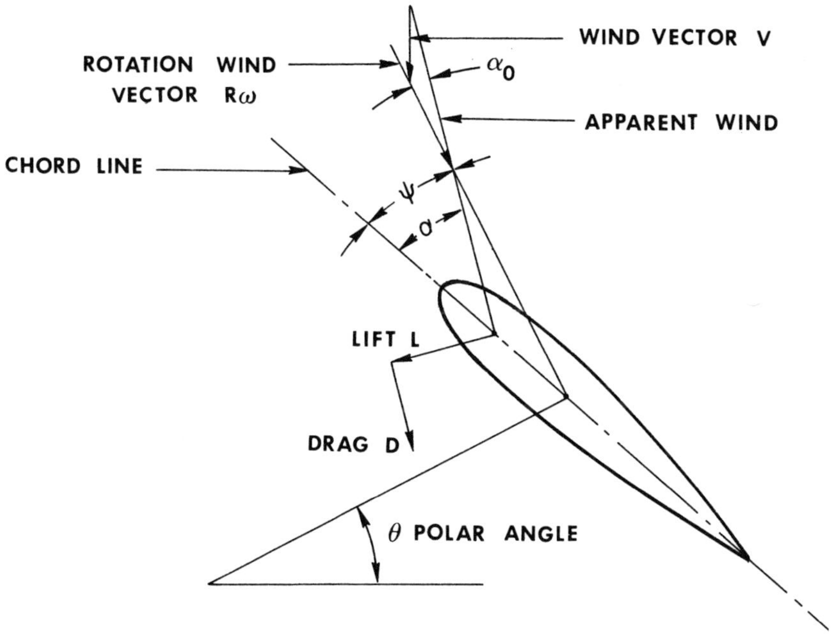

Figure 2. Principal flow relations for an element of a blade

The figure indicates a pitch angle ψ which would be variable in an articulated blade giromill turbine, but would vanish for Darrieus rotors. The variation of the relative flow about a blade section in the four quadrants of the cycle is shown in Figure 3 for the special case of identical local streamline velocity V_s. Actual velocities will vary considerably with the polar angle θ, particularly in the leeward part of the cycle.

If a reasonable model could be developed for the distribution of longitudinal and transverse interference factors then stream surface analyses connecting the forces on the blades with defects in momentum in the wake would probably yield reasonable approximations for the power coefficients. As an indication of the approach which might be taken in the analysis some relations will be presented here for the situation of constant local streamline velocity.

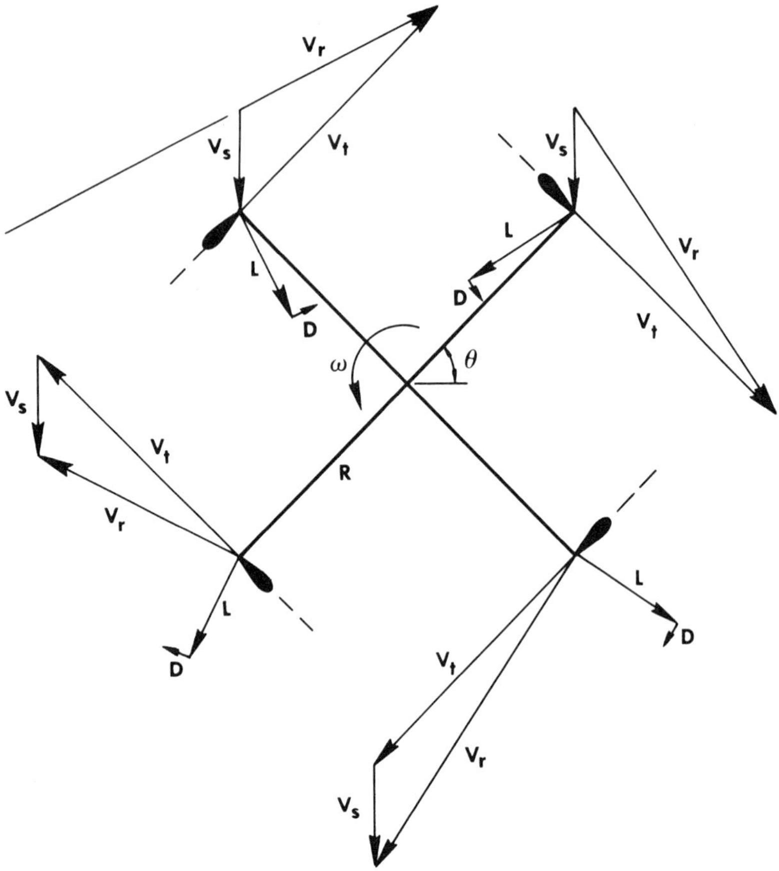

Figure 3. Local flow over a blade section in the four quadrants of the cycle

The unit vector, T, along the tangent to the blade motion is

$$T = (-\sin\theta, \cos\theta) \quad (2)$$

The absolute wind vector, V_s, is

$$V_s = V_s(0, -1) \quad (3)$$

The relative wind vector is

$$V_t = R\omega(\sin\theta, -\cos\theta) \quad (4)$$

The apparent wind vector is therefore

$$V_r = (R\omega\sin\theta, \; -(V_s + R\omega\cos\theta))$$

$$= V_s w(x\sin\theta, \; -(1 + x\cos\theta)) \qquad (5)$$

$$w = ((x\sin\theta)^2 + (1 + x\cos\theta)^2)^{\frac{1}{2}}$$

with

$$x = R\omega/V_s$$

With zero pitch angle the angle of attack α is given by

$$\sin\alpha = T \times V_r/V_s w = -\sin\theta/w \qquad (6)$$

Lift coefficient is $C_L = C_{L,\alpha}$ and the lift vector is given by

$$L = L_0(-(1 + x\cos\theta), \; -x\sin\theta) \qquad (7)$$

with

$$L_0 = \rho c V_s^2 \, w^2 \, C_{L,\alpha} \sin\alpha/2$$

Here, c is the local blade chord and lift is on a unit length of span which in the following analysis will be assumed to be parallel to the axis. The drag coefficient for wings is given by

$$C_D = C_{D,0} + C_L^2 / \pi A_R \qquad (8)$$

which is the sum of form and induced drag coefficients. The latter is rather difficult to determine for wind turbine blades, particularly in the case of the curved Darrieus rotors. The torque is given by the summation of the effects of lift L and drag D. The instantaneous power P is the product of the torque and the angular rate. The instantaneous power coefficient c_P defined on the projected swept area is given by

$$c_P = \frac{\omega(L\cos(L,T) - D(\cos(D,T))}{2R\rho V_s^3 / 2} \qquad (9)$$

For N blades the instantaneous power coefficient should include the phase relations of the blades but since an average power coefficient is of ultimate interest, Equation 9 may be simplified to the form

$$c_P = \frac{Nc}{2R} Xw\left[C_L \sin\theta - \left(C_{D,0} + C_L^2 / \pi A_R\right)\left(x + \cos\theta\right)\right] \qquad (10)$$

with

$$X = R\omega/V, \text{ tip speed ratio}$$

Power coefficient for the machine is the average of local instantaneous power coefficients and is given by

$$C_P = \frac{Nc}{2R}\frac{X}{2\pi}\int_0^{2\pi} w\left[C_L \sin\theta - \left(C_{D,0} + C_L^2 / \pi A_R\right) \cdot \left(x + \cos\theta\right)\right] d\theta \tag{11}$$

This relation was used to calculate the performance of a vertical axis wind turbine with the following parameters:

$C_{L,\alpha} = 4.0$ $\quad C_{D,0} = 0.01$ $\quad \pi A_R = 30$

$Nc/2R = 0.075$

The interference factor was taken as 1/3, the optimum value for propeller turbines so that $x = 1.5X$. Slope of the lift coefficient, $C_{L,\alpha}$ was taken as approximately 2/3 of the ideal value to account for the curvature of the Darrieus rotor. If the tip speed ratio X is reduced by about 25% to account for the variation of radius then the results, (Figure 4) are quite similar to those shown in Figure 1. It is highly questionable whether detailed calculations for blade elements with accounting for the inclination to the axis of rotation and local radius are worthwhile until rational models for interference factors are developed.

It is clear from the above that the current state of analytical techniques in vertical axis wind turbine engineering is far from satisfactory. Small scale model tests are of questionable value because aerodynamic performance of the blades deteriorates at low Reynolds numbers. The remaining alternative of full scale testing is so expensive that only relatively slow development in this field may be expected.

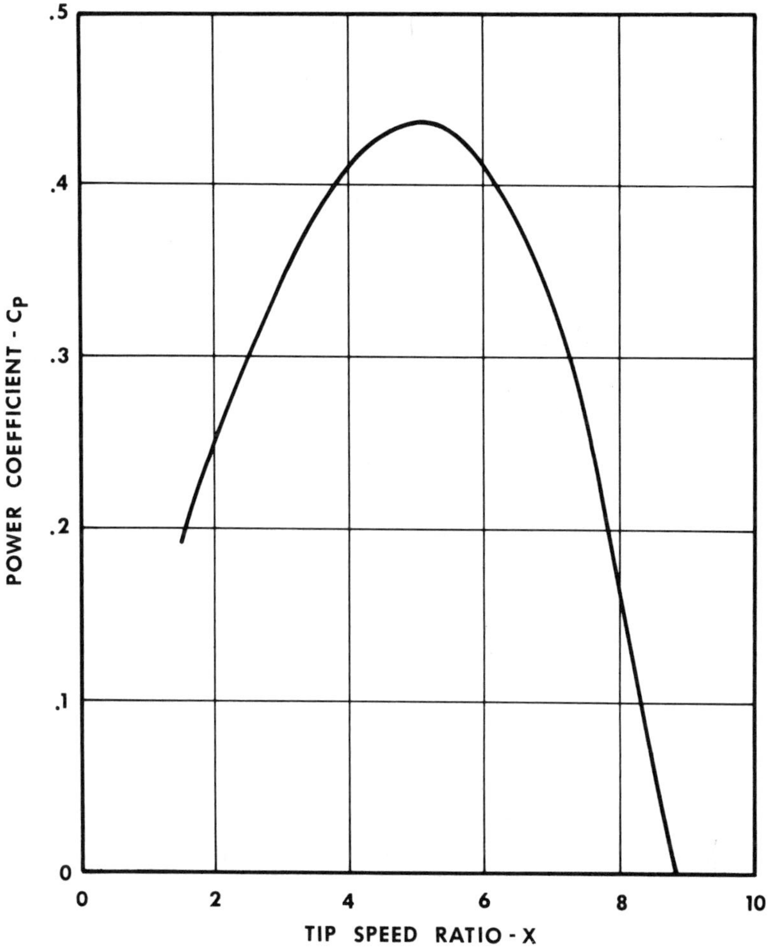

Figure 4. Calculated performance of a vertical axis wind turbine

REFERENCES

1. Wilson, R. E., Walker, S. N., and Lissaman, P. B.S., "Aerodynamics of the Darrieus Rotor," AIAA Journal of Aircraft, Vol. 15, No. 12, Dec. 1976, pp. 1023-1024.
2. Templin, R. J., "Aerodynamic Performance Theory for the NRC Vertical-Axis Wind Turbine," National Research Council of Canada, LTR-LA-160, June 1974.
3. Muraca, R. J., Stephen, S., Maria, V., and Dayenhart, R. J., "Theoretical Performance of Vertical Axis Windmills," NASA TMX-726 662, May 1975.
4. Sheldahl, R. E., Klimas, P. C., and Feltz, L. V., "Aerodynamic Performance of a 5-M-Diameter Darrieus Turbine," AIAA Journal of Energy, Vol. 4, No. 5, Sept-Oct. 1980, pp. 227-232.
5. Ashley, H., "Some Contributions to Aerodynamic Theory for Vertical Axis Wind Turbines," AIAA Journal of Energy, Vol. 2, No. 2, March-April 1978, pp. 113-119.

CHAPTER 6

DRAG-TYPE TURBINES

Drag type wind turbines which depend on the differences of the wind forces on the arms which are advancing into and retreating from the wind are among the earliest wind machines developed by man. In the earliest machines, found extensively in ancient Persia, the advancing arms were shielded from the wind by an enclosure while the retreating arms were fully exposed. The principal problem with such devices was their fixed orientation which did not allow them to take full advantage of winds coming from all directions.

Recent efforts in drag turbines have concentrated on articulated drag surfaces which expose maximum area when retreating and attempt to minimize drag on the advancing part of the cycle by reducing their frontal area. This is usually achieved by means of balanced shutters which close when the relative wind is from one direction and open fully when it is from the other. In the case of paddles on the ends of arms, these are usually rotated to align them with the wind vector at the beginning of the advance into the wind.

A relatively recent variant of the drag-type wind turbine is the Savonius rotor which consists of an S-shaped surface whose rotary motion is maintained by the torque difference between the retreating concave arm and the advancing convex member.

Regardless of the details of construction, all drag-type wind turbines suffer from relatively low efficiencies and bulky, unattractive appearance. Their one outstanding characteristic is the very high starting torque which is lacking in propeller and Darrieus turbines.

PADDLE TURBINES

There are no satisfactory theories for the operation of paddle turbines, particularly if a dense array of paddles is employed. A first order of magnitude for machines with up to four arms may be obtained from very simple analysis in which it is assumed that the local drag of each element of the paddle at a particular radius from the axis is due only to local normal component of the relative wind speed.

Initially attention will be focused on a single paddle of height h and radial extent ℓ with ℓ being so much smaller than the radius of the arm R that radial variations of relative speeds on the paddle may be neglected. It will also be assumed that at the moment that the arm is pointing into the wind it is instantaneously rotated to present the total paddle area ℓh to the wind. When advancing into the wind, the paddle is feathered and presents a relatively low resistance. The basic relations are shown in Figure 1.

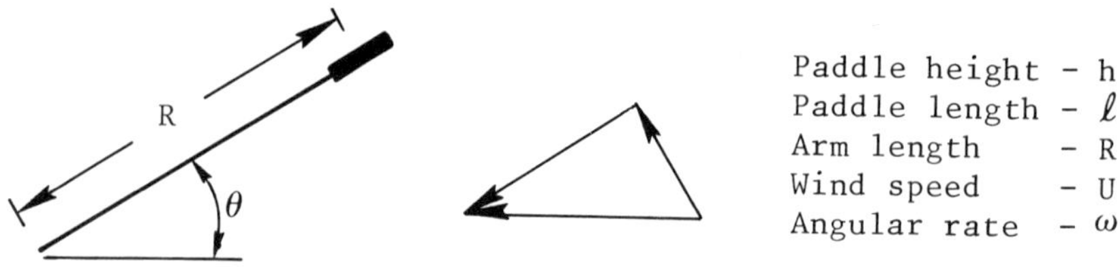

Figure 1. Definition of the paddle turbine parameters

In the simple normal flow impact theory of drag, the drag coefficient is assumed to vary as the square of the sine of the angle between the surface and the oncoming stream. Hoerner, in Fluid Dynamic Drag (1) shows that the drag coefficient relation

$$C_D = C_{D,0} \sin^2\theta \qquad (1)$$

gives excellent agreement with test results over a very large range of angles θ. In this case, the paddle is retreating from the normal component of the wind, $U\sin\theta$, at a rate $R\omega$ so that the drag is given by

$$D = (\rho/2)C_{D,0}\, \ell h(U\sin\theta - R\omega)^2 \qquad (2)$$

with ρ being the density of air. The torque is given by $T = RD$, and instantaneous power $P = T\omega$. Instantaneous power coefficient $c_p(\theta)$ is given by

$$c_p(\theta) = \frac{P}{qU2Rh} = fX(\sin\theta - X)^2 \qquad (3)$$

with q being the dynamic pressure, $X = R\omega/U$, the tip speed ratio, and $f = C_{D,0}\ell/2R$, a geometrical parameter. It is assumed that when the arm begins to advance into the wind, the paddle is feathered so that its drag coefficient is $C_{D,1}$ and the geometrical parameter becomes f'. The average power coefficient over a complete cycle for one paddle is

$$C_{P,1} = \frac{X}{2\pi}\left[f\int_0^{\pi}(\sin\theta - X)^2 d\theta - f'\int_{\pi}^{2\pi}(\sin\theta - X)^2 d\theta\right]$$

$$= (f - f')\left[\frac{X}{4} - \frac{2\Delta}{\pi}X^2 + \frac{X^3}{2}\right] \qquad (4)$$

with

$$\Delta = \frac{f + f'}{f - f'}$$

The maximum value of $C_{P,1}$ is achieved at

$$X = \frac{4\Delta}{3\pi} - \left[\left(\frac{4\Delta}{3\pi}\right)^2 - \frac{1}{6}\right]^{\frac{1}{2}} \tag{5}$$

With f = 0.07 and f' = 0.01 the maximum power coefficient is found to be 0.00123 at X = 0.174. The average power coefficient becomes negative at X = 0.39. Even with four paddles and no interference effects the maximum power coefficient is less than 1.0% of the theoretical value which can be achieved with propeller turbines.

A variation of the above concept is found in the case of a paddle which through linkages is made to always be normal to the oncoming wind. With the same impact theory assumptions as were used above the instantaneous power of the machine is

$$P_{(\theta)} = qU\ell hC_{D,0}\, X\sin\theta(1 - X\sin\theta)^2 \tag{6}$$

with the $\sin\theta$ term outside the parenthesis coming from the effective moment arm which is $R\sin\theta$. If on the advancing part of the cycle the drag coefficient is $C_{D,1}$, then the average power coefficient is

$$C_{P,1} = \frac{\int_0^{2\pi} Pd\theta}{2\pi qU2Rh} = (f - f')\left[X - \frac{\pi\Delta}{2} X^2 + \frac{2}{3} X^3\right] \tag{7}$$

with the definitions of f, f', and Δ remaining unchanged. The maximum value of $C_{P,1}$ is found at

$$X = \frac{\pi\Delta}{4}\left[1 - \left(1 - 8/(\pi\Delta)^2\right)^{\frac{1}{2}}\right] \tag{8}$$

As was done previously, it is assumed that f = 0.07 and f' = 0.01 which leads to a maximum power coefficient of .0078 at X = 0.275. This illustrates the futility of this particular articulation scheme, and all its related notions which attempt to maintain drag surfaces at normal incidence to the nominal wind direction.

A more elaborate form of paddle-type turbine is a drag surface composed of balanced shutters which close under wind pressure from one side and open fully, when the direction of the pressure is reversed. Numerous variants of such schemes have been proposed and they all share the problems of low efficiency of such devices. It is generally not realized that when shutters are actuated by wind pressure useful power is generated over a very small portion of each cycle. This can be shown in very simple analysis similar to those presented above.

The shutters will close only when the net force on the shutter is in the direction of rotation during the time that the arm is retreating from the wind. At all other times, and particularly during the advance into the wind, the shutters stay open. As before, a simple impact theory will be used to estimate aerodynamic forces. It should be noted that when $r = (U\sin\theta)/\omega$ the relative wind normal flow component vanishes. Thus, positive pressure which tends to close the shutters exists only up to that radius. With pressure given by

$$p = \rho v_n^2/2 = q(\sin\theta - x)^2 \qquad (9)$$

with q being the free stream dynamic pressure, $\rho U^2/2$, and $x = r\omega/U$ the local tangential speed ratio. The closing force is therefore

$$T^+ = \frac{qRH}{X}\int_0^{x_0 = \sin\theta} (\sin^2\theta - 2x\sin\theta + x^2)dx$$

$$= \frac{qRH}{X}\frac{\sin^3\theta}{3} \qquad (10)$$

where $X = R\omega/U$ the tip speed ratio. The force tending to open the shutters is the integral of the force from x_0 to X.

$$T^- = \frac{qRH}{X}\left(X\sin^2\theta - X^2\sin\theta + \frac{X^3}{3} - \frac{\sin^3\theta}{3}\right) \qquad (11)$$

When these two forces are equated

$$Y^3 = \frac{3}{2}(Y^2 - Y) + \frac{1}{2} \qquad (12)$$

with

$$Y = \sin\theta/X$$

The relevant solution of the above equation is Y = 1/2 which means that the shutters will remain closed when

$$\sin\theta \geqslant X/2 \qquad (13)$$

and useful power will be generated. Power output can be calculated by integrating the torque over the arm taking into account the fact that the outer regions will be making negative contributions until $\sin\theta > X$. As a simple approximation, the power coefficient may be estimated by considering only the power generated when only positive pressures exist over the whole paddle. Element of torque is

$$dT = qR^2H\left(\sin^2\theta - 2Xt\sin\theta + X^2t^2\right)tdt \tag{14}$$

with

$$t = r/R$$

Integration yields

$$T = qR^2H\left(\frac{\sin^2\theta}{2} - \frac{2X\sin\theta}{3} + \frac{X^2}{4}\right) \tag{15}$$

Instantaneous power coefficient, defined on 2RHqU is

$$c_p(\theta) = \frac{X}{2}\left(\frac{\sin^2\theta}{2} - \frac{2X\sin\theta}{3} + \frac{X^2}{4}\right) \tag{16}$$

which when integrated over the interval

$$\theta_1 = \arcsin X \leqslant \theta \leqslant \pi - \theta_1 \tag{17}$$

and averaged over the full cycle of 2π radians results in the power coefficient for 1 paddle which is given by

$$C_{P,1} = \frac{X}{4\pi}\left[\left(1 + X^2\right)\left(\frac{\pi - 2\theta_1}{4}\right) + \frac{\sin 2\theta_1}{4} - \frac{4X\cos\theta_1}{3}\right] \tag{18}$$

Calculated values of $C_{P,1}$ are shown in Figure 2. The maximum value of .012 which occurs at X = 0.45 is very low. Mutual interference effects would appear with 4 paddles so that a value of 0.048 probably could not be achieved even theoretically. With full accounting for the induced flow interference and wake effects, it is quite unlikely that the above figures could be approached even remotely.

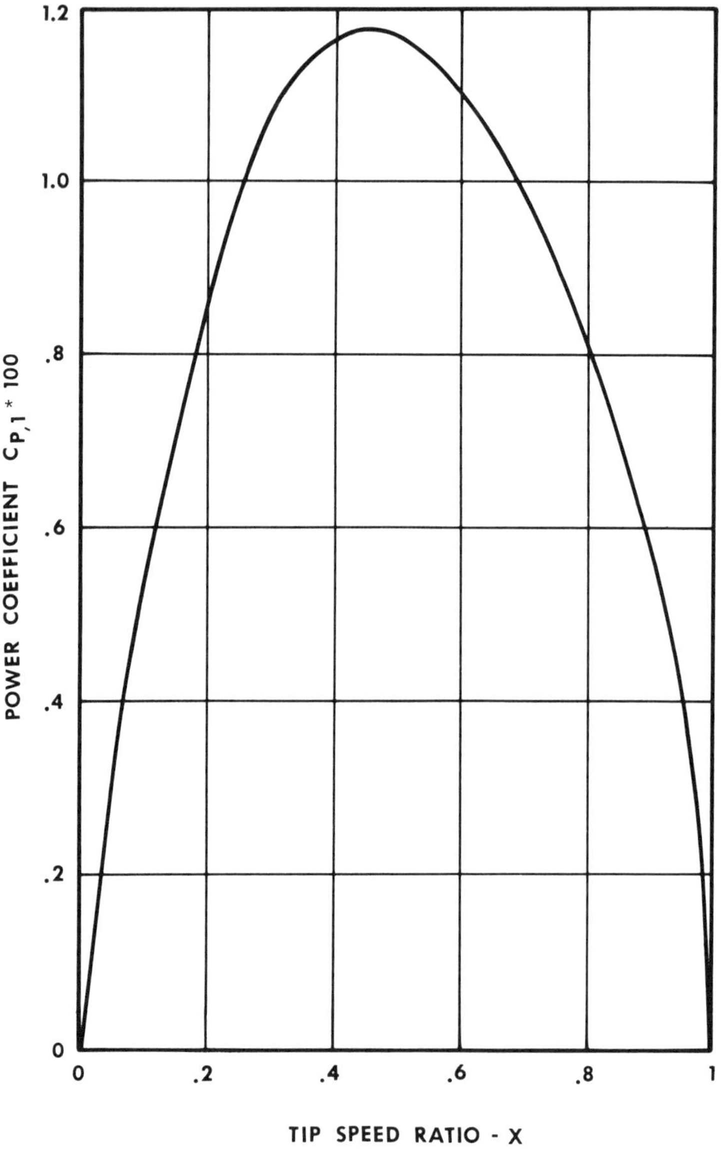

Figure 2. Power coefficient of shutter type paddle wheel wind turbine - output of one paddle

SAVONIUS ROTORS

Savonius rotors are drag type wind turbines which depend primarily on the difference in drag of curved advancing and retreating elements to generate torque. As in all fields, numerous variants have been developed, but the principal types are the single and split rotor configurations which are shown schematically in Figure 3

a. Split Rotor b. Single Rotor

Figure 3. Principal types of Savonius rotors

Also shown in the Figure are baffles, or augmenting devices which are frequently employed to increase the power output of the basic rotor. Common to all the designs is a large rotating surface which has a very high drag/lift ratio, and therefore an inherently low efficiency of extraction of power from the air stream. The principal advantage of these turbines lies in their rugged simplicity, low cost, easy fabrication and insensitivity to changes in wind directions. Proponents of Savonius rotors advance the above arguments to counter the objections of large unsightly devices which produce relatively little power.

The complex nature of the flow around Savonius rotors has discouraged attempts at analyses like those developed for propeller wind turbines. Unlike high speed propellers Savonius rotors are extremely close-coupled configurations in which the blades are in very close proximity to each other and present a very large time dependent projected area to the flow. Both the advancing and retreating blades shed vortices into the wake which is strongly influenced by the induced circulation around the rotor. Efforts to analyze the performance of Savonius rotors in terms of equivalent actuator disks corresponding to propellers must be dismissed as totally unrealistic assumptions which have no basis in fact and no hope for success.

Numerous experimental studies of large and small models of Savonius wind turbines (e.g., Sabzevari Reference 1),

Charwat (Reference 2) have yielded engineering results without adding significantly to the understanding of the problem. However, sufficient data exist to indicate the characteristics of the performance curves, i.e., variation with the speed of rotation of the torque coefficient

$$C_T = \frac{T}{\rho U^2 Ad/2}$$

where T is torque, A the projected area, and d the diameter. It has been found that in the middle of the range of operating tip speed ratios of simple and augmented Savonius rotors the variation of the torque coefficient with tip speed ratio is almost linear. This is shown in Figure 4 which is taken from Reference 3.

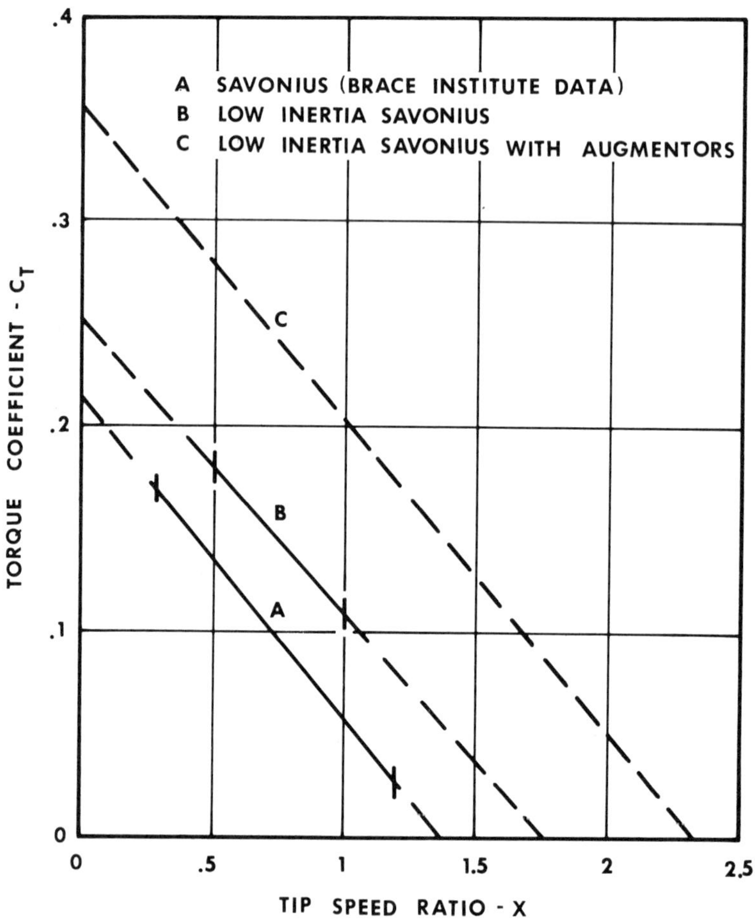

Figure 4. Performance of representative Savonius rotors, including the effects of augmentation

The relatively high tip speed ratios appearing with augmented rotors are due to the use of the basic wind speed for reference, even though the rotor is operating in a flow accelerated by the augmenting surfaces. The torque coefficient may be approximated by

$$C_T = C_{T,0} - kX \qquad (19)$$

where $C_{T,0}$ is the stall torque coefficient and k is the constant of proportionality which is equal to $C_{T,0}/X_T$ with X_T being the limiting tip speed ratio. The power coefficient is written as

$$C_P = 2XC_T = 2C_{P,0}X(1 - X/X_T) \qquad (20)$$

Maximum value of the power coefficient is achieved at X/X_T = 1/2 and is equal to $C_{T,0}\ X_T/2$. It is more convenient to normalize on the design tip speed ratio X_0 which corresponds to the maximum power coefficient, $C_{p,m}$ so that

$$C_P = 2C_{P,0}\dot{W}(1 - \dot{W}/2) \qquad (21)$$

with

$$\dot{W} = X/X_0$$

The curves in Figure 4 all have slopes equal to about 0.45 and this is also approximately true of the single rotor curves presented by Charwat (Reference 2). This remarkable fact leads to considerable simplifications in making engineering estimates since performance of Savonius rotors is approximated by a function of X with only $C_{T,0}$ as a parameter. For preliminary design estimates it is therefore possible to generate a set of one parameter curves of the power coefficient

$$C_P = 2C_{T,0}\ X(1 - kX/C_{T,0}) \qquad (22)$$

with k = 0.15, and

$$X_0 = C_{T,0}\ /2k$$

$$C_{P,0} = C_{T,0}^2\ /2k$$

Calculated values of the power coefficient are shown in Figure 5. Also shown in the Figure is the locus of the maxima of the power coefficient. With the above definitions the power coefficient relation may be rewritten as

$$C_P/C_{P,0} = 2(X/X_0)\left[1 - (X/X_0)/2\right] \qquad (23)$$

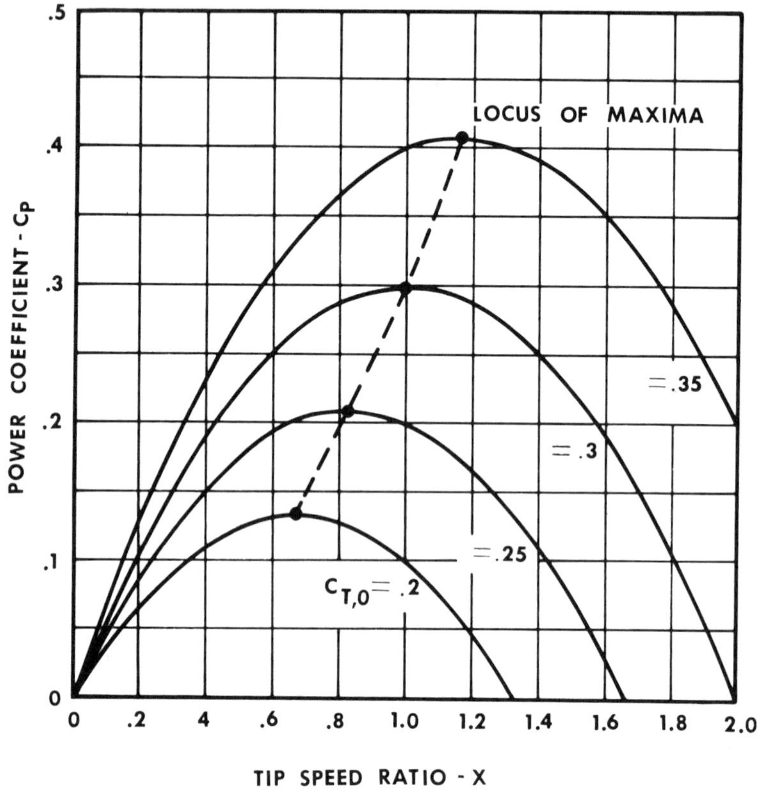

Figure 5. Power coefficients of augmented Savonius rotors

In certain circumstances it is advantageous to offset the design tip speed ratio X_d from the optimum value X_0 in order to maximize the capture of wind energy over a period of time. If X_d corresponds to a design angular rate ω_d and a nominal wind speed V_0 then at wind speed V the tip speed ratio X is given by

$$X = \frac{\omega_d D/2}{V} = \frac{\omega_d D/2}{V_0} \frac{V_0}{V} = X_d \frac{V_0}{V_a} \frac{V_a}{V} \qquad (24)$$

Here the average or reference wind speed V_a is introduced since most annual wind speed distributions are expressed in terms of V/V_a. The power coefficient therefore becomes

$$C_P = C_{P,0} \frac{4Y}{\lambda} (1 - Y/\lambda) \qquad (25)$$

with

λ = wind speed ratio V/V_a

V_a = reference wind speed for a given site

V_0 = design wind speed

$Y = (X_d V_0 / X_0 V_a)/2$

In the above relation, X_0 characterizes the machine, X_d is the design tip speed which may differ from the design value, V_0/V_a is the ratio of the design speed to the nominal wind speed and $\lambda = V/V_a$ is the wind speed ratio which is a function of time, expressed in probability from.

The above relations are true for augmented and basic Savonius rotors which are characterized by the stall torque coefficient $C_{T,0}$. Operation at design conditions X_d, which correspond to the characteristic tip speed ratio X_0 and design wind speed equal to the nominal wind speed implies $Y = 1/2$. Any other value implies that the design of the machine is biased in some way to maximize the total energy extraction. Variation of the power coefficient with the wind speed ratio is shown in Figure 6.

A SIMPLE SAVONIUS ROTOR FLOW ANALYSIS

The complexity of the flow in and around a Savonius rotor precludes detailed analysis at this point. It is possible, however, the derive relatively simple engineering relations from basic analyses of simple flow models. An example is presented here to illustrate a possible approach to the problem.

A Savonius rotor induces a rotary motion in the flow ahead of it while at the same time presenting a bluff, roughly cylindrical obstacle to the oncoming stream. At the rear of the rotor where the flow is discharged the main stream separates and a low energy wake is formed. The flow ahead of the rotor is essentially inviscid and can be modelled quite easily. In the wake region, it is necessary to make broad assumptions regarding the nature of the flow and to argue in terms of overall momentum and energy balances. The simplest hydrodynamic model which exhibits flow blockage and induced circulation is a doublet with a superimposed vortex. The complex potential for a uniform flow in the $-\pi/2$ direction, with a doublet opposing it at the origin and a vortex also located at the origin is given by (e.g., Streeter, Reference 4)

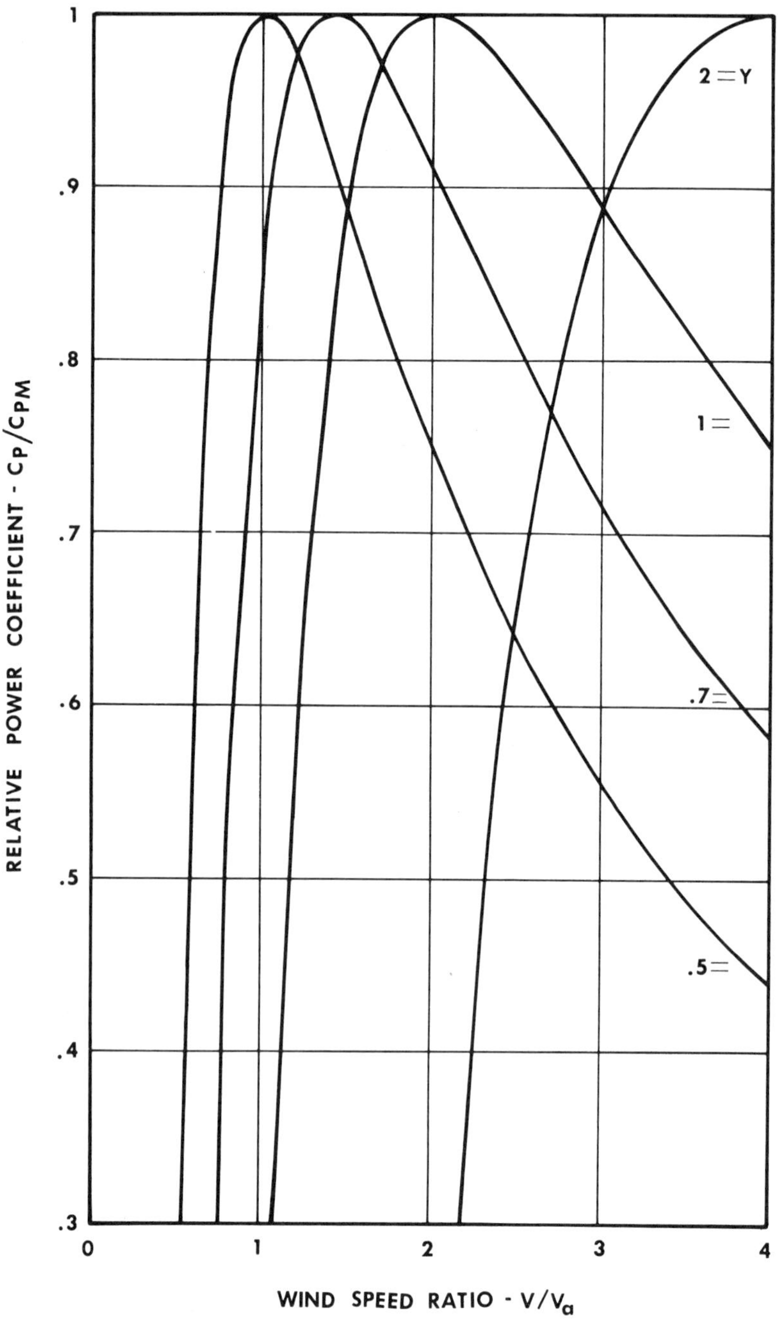

Figure 6. Variation of the relative power coefficient with the wind speed ratio

$$W = \phi + i\psi = -iUz + iK\ln z + i\mu/z \qquad (26)$$

Here ϕ and ψ are the potential and stream functions respectively, U is the free stream velocity, $2\pi K$ the circulation of the vortex and μ the strength of the doublet. In polar coordinates with radius r and angle θ

$$\phi = rU\sin\theta - K\theta + \mu\sin\theta/r \qquad (27a)$$

$$\psi = -rU\cos\theta + K\ln r + \mu\cos\theta/r \qquad (27b)$$

The radial and tangential components of velocity, v_r and v_θ respectively, are given by

$$v_r = -\frac{\partial\phi}{\partial r} = -\frac{1}{4}\frac{\partial\psi}{\partial\theta} \qquad (28a)$$

$$v_\theta = -\frac{1}{r}\frac{\partial\phi}{\partial\theta} = \frac{\partial\psi}{\partial r} \qquad (28b)$$

It is convenient to put the equations in dimensionless form by defining

$$X = \frac{K}{RU} \qquad a = \frac{\mu}{UR^2} \qquad \lambda = r/R$$

so that

$$v_r/U = -\left(1 - a\lambda^{-2}\right)\sin\theta \qquad (29a)$$

$$v_\theta/U = -\left[\left(1 + a\lambda^{-2}\right)\cos\theta - X\lambda^{-1}\right] \qquad (29b)$$

The parameter X is the ratio of the tangential velocity at the rotor radius R to the free stream velocity, and thus is the usual tip speed ratio. The parameter a is the blockage of the flow since the flow rate into the rotor at radius R is given by

$$\dot{m} = -\int_0^\pi \rho v_r R d\theta = \rho UR(1 - a)\int_0^\pi \sin\theta d\theta$$

$$= 2\rho UR(1 - a) \qquad (30)$$

Representative streamline patterns for a = .2 and X = 0.85 are shown in Figure 7.

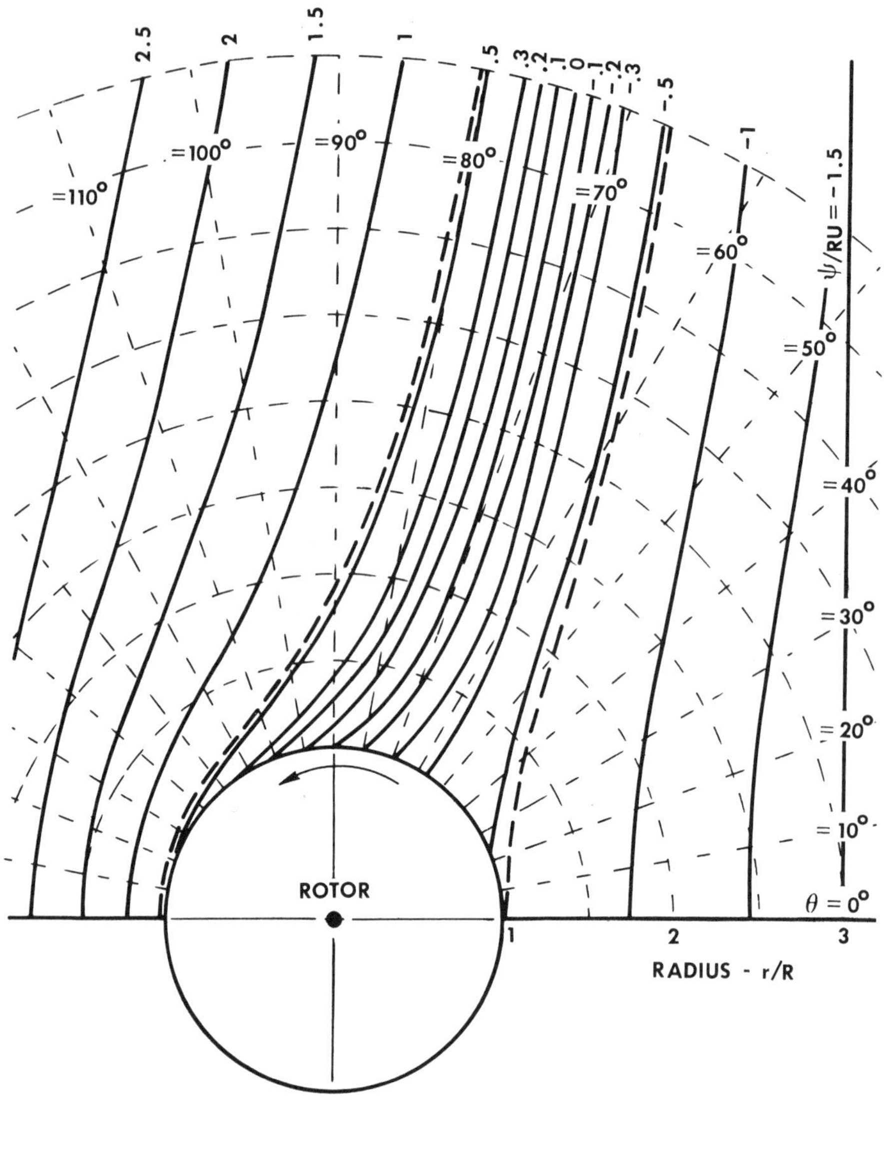

Figure 7. Streamlines of flow at a Savonius rotor with a = 0.2 and X = 0.85

At very large distances ahead of the rotor the velocity is U, but at the rotor it is

$$V = \left(v_r^2 + v_\theta^2\right)^{\frac{1}{2}} = U\Big[(1 - a)^2 \sin^2\theta +$$

$$+ (1 - a)^2 \cos^2\theta - 2X(1 + a)\cos\theta + X^2\Big] \qquad (31)$$

Since no energy is gained or dissipated ahead of the rotor the Bernoulli equation is written as

$$p = p_\infty + \rho\left(U^2 - V^2\right)/2 \qquad (32)$$

The axial thrust on the control volume at the forward half of the rotor is therefore calculated to be

$$T = \int_0^\pi pR\sin\theta d\theta$$

$$= 2R\left\{p_\infty + \frac{\rho U^2}{2}\left[\frac{2}{3}(1 - a)^2 + \frac{1}{3}(1 + a)^2 + X^2\right]\right\}$$

$$(33)$$

The average pressure coefficient on the control surface at the forward half of the rotor is therefore written as

$$\frac{p_i - p_\infty}{\rho U^2/2} = 1 - \left[\frac{2}{3}(1 - a)^2 + \frac{1}{3}(1 + a)^2 + X^2\right]$$

$$(34)$$

After exit from the rotor the flow has some exit velocity u_e and a tangential component u_θ, with the average pressure p_b. Very far downstream the axial velocity is taken to be $U - \Delta U$, a residual transverse velocity $\sqrt{e}\, u_\theta$ is assumed to be present and the pressure is taken to be p_∞. Since no work or dissipation is assumed to take place between the control surface at the base and a corresponding one far downstream the Bernoulli equation is written as

$$\frac{p_b - p_\infty}{\rho U^2/2} = (1 - Q)^2 - \left(w_e^2 + X^2\right) + eX^2 \qquad (35)$$

with

$$Q = \Delta U/U$$

In the above relation the tangential velocity at the base control surface is taken to be equal to the tip speed of the rotor. The net pressure coefficient on the rotor is now written as

$$\frac{p_i - p_b}{\rho U^2/2} = Q(2 - Q) - \frac{2}{3}(1 - a)^2 - \frac{1}{3}(1 + a)^2 + w_e^2 - eX^2 \tag{36}$$

The overall momentum balance on the flow going through the rotor must include the changes in momentum as well as the pressure and rotor thrust, T_1' on the fluid. The balance of forces on the fluid is written as

$$\dot{m}U(1 - Q) - \dot{m}U = \left(p_i - p_\infty\right)a_1 + \int_{S_1}\left(p - p_\infty\right)dS_1 + \left(p_b - p_\infty\right)2R(1 - X) + \int_{S_2}\left(p - p_\infty\right)dS_2 - \left(p_2 - p_1\right)a_2 + T' \tag{37}$$

In the above

a_1, a_2	are the stream tube areas very far ahead and downstream of the rotor
p_1, p_2	are pressures very far ahead and downstream of the rotor
$p_b - p_\infty$	is the net pressure on the stream surface in the base region
$\int_{S_1}, \int_{S_2}$	are integrals of the pressure forces on the stream tube ahead and downstream of the rotor

Following the usual arguments from propeller theory the pressure terms may be neglected with the exception of the base pressure term. The fraction of the base area over which the base pressure acts on the stream tube may be estimated from the continuity equation since

$$2R(1 - a)U = XU_e \tag{38a}$$

or

$$1 - X = 1 - (1 - a)/w_e \tag{38b}$$

The coefficient of thrust on the rotor is now written as

$$\frac{T}{2R\rho U^2/2} = 2(1 - a)Q + C_B\left[1 - (1 - a)/w_e\right] \tag{39}$$

with

$$C_B = 2\left(p_b - p_\infty\right)/\rho U^2$$

When the above is equated to the coefficient given by Equation 36 the velocity deficit $Q = \Delta U/U$ is found to be given by

$$Q = a + \left\{w_e^2 - C_B\left[1 - (1 - a)/w_e\right] - 1 - \right.$$
$$\left. - eX^2 + 2a/3\right\}^{\frac{1}{2}}$$

The above relation may be simplified considerably when the quantities in the second term are estimated at the bounding values of a. When the rotor presents a minimum resistance to the flow a tends towards zero and w_e tends to unity. As a first approximation for small values of eX^2

$$w_e = 1 + eX^2(1 + C_B/2) \tag{41}$$

When the rotor is nearly stopped it presents the maximum resistance to the flow so that a tends to unity and w_e vanishes. In order to avoid singularities w_e must vanish in proportion to $(1 - a)$. Further, in order to avoid complex variables it is necessary to impose the condition

$$w_e \geqslant \frac{1 - a}{1 - (1/3 + eX^2)/(-C_B)} \tag{42}$$

It is reasonable to assume that w_e varies directly with (1 - a) so that when $a \rightarrow 1$ $Q \rightarrow a$. For low values of a, corresponding to the minimum obstruction of the flow the second term in Equation 40 also tends to zero. As a first approximation the velocity defect in the wake will be taken as

$$Q = a \tag{43}$$

Power output is the change in the kinetic energy of the flow going through the machine, or

$$P = \dot{m}\left[U^2 - U^2(1 - Q)^2 - eX^2U^2\right]/2 \tag{44}$$

In terms of the power coefficient C_P the performance of Savonius rotors is expressed by

$$C_P = \frac{P}{2R\rho U^2/2} = (1 - a)\left[2a - a^2 - eX^2\right] \quad (45)$$

It is more convenient to operate in terms of the tip speed ratios rather than interference factors. In this case the two are related since the flow through the machine may be written as

$$2R(1 - a)U = 2RC_F UX \quad (46)$$

which expressed the volume flow rate in terms of the rotational speed of the machine and a flow coefficient C_F. If at the terminal tip speed ratio X_T the flow rate is a fraction f of the maximum value 2RU then

$$1 - a = fX/X_T \quad (47)$$

With this relation the power coefficient may be written as

$$C_P = f\frac{X}{X_T}\left[1 - (f^2 + e)\left(\frac{X}{X_T}\right)^2\right] \quad (48)$$

Since it was assumed that the terminal value of the tip speed ratio was X_T

$$e = (1 - f^2)/X_T^2 \quad (49)$$

and

$$C_P = f\frac{X}{X_T}\left[1 - \left(\frac{X}{X_T}\right)^2\right] \quad (50)$$

The maximum value of the power coefficient occurs at

$$\left(X/X_T\right)_{opt} = (1/3)^{\frac{1}{2}} = 0.577 \quad (51)$$

and is equal to

$$C_{P,0} = 2f/3^{1.5} = .385f \quad (52)$$

Variation of the power and torque coefficients with X is shown in Figure 8 for f = 1.0 and X_T = 1.5. The shape of the power coefficient curve looks quite reasonable and in the mid range the torque coefficient could be approximated by a straight line.

In view of the numerous simplifying assumptions contained in the analysis it is not possible to present conclusive arguments regarding the functional relationships for the performance curves, and the derived equations should be viewed as engineering approximations

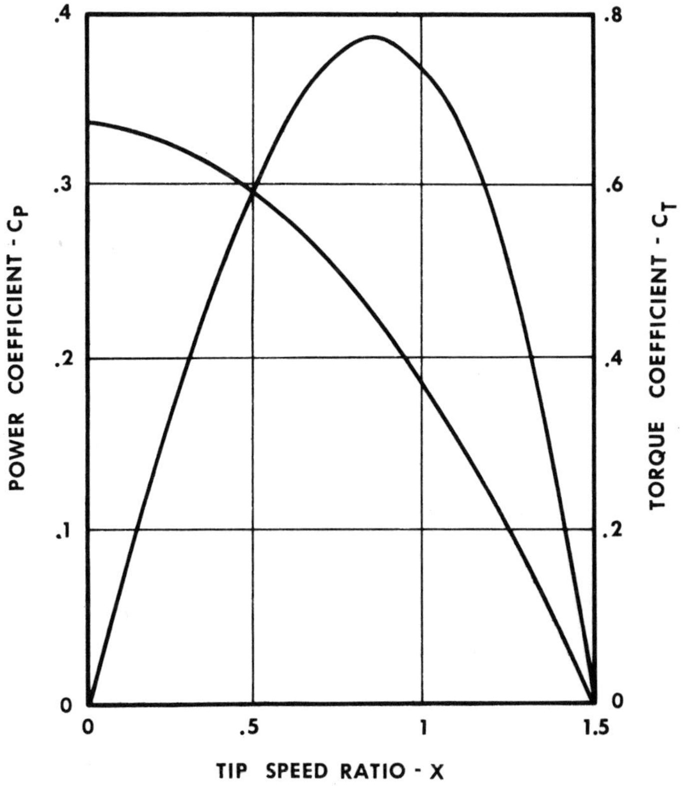

Figure 8. Power and torque coefficient curves for a Savonius rotor with f = 1 and X_T = 1.5

REFERENCES

1. Sabzevari, A., "Performance Characteristics of Concentrator Augmented Savonius Wind Rotors," Wind Engineering, Vol. 1, No. 3, 1977, pp. 198-206.
2. Charwat, A. F., "Performance of Counter- and Corotating Arrays of Savonius Turbines," AIAA Journal of Energy, Vol. 2, No. 1, Jan-Feb 1978, pp. 61-62.
3. Wortman, A., "Guidelines for Evaluating Wind Turbine Generating Systems," AWD Inc., Santa Monica, CA, Report AWD-R-81-0201, Feb. 1981.
4. Streeter, V. L., Fluid Dynamics, McGraw-Hill Book, Co., New York, N.Y., 1948.

CHAPTER 7

EXTRACTION OF ENERGY FROM THE WIND

Performance of wind turbines in given wind conditions can be readily determined in engineering tests or estimated using the available calculational techniques. Some of these were outlined in the preceding chapters to indicate the current state of the art and the uncertainties involved in all the analytical approaches. In order to be able to estimate the energy capture capabilities of a wind turbine over long periods of time when a wide range of wind conditions is encountered, it is necessary to determine the turbine operating characteristics or performance curves. Wind tunnel testing of scale models is of relatively small value because of the Reynolds number effects so that it is usually necessary to undertake full scale testing. Even then, the results must be treated with some caution because the unsteady nature of the wind involves the transient response of the wind turbine system which cannot be determined from the time integrals of wind speed and power outputs. Thus, an unknown function, which is certainly site dependent, remains as an undetermined parameter of the problem.

Preliminary engineering and economic feasibility studies are based on the value of the energy captured in a wind turbine lifetime relative to the initial investment and operating and operating and maintenance costs. The total wind energy available depends on the location and may have a significant annual variation. Detailed data for the distribution of wind speeds and directions over a period of several years are available for a very limited number of sites so that the statistical data base for long term predictions of energy capture is rather sparse. It is generally necessary to extrapolate from limited quantities of data and to assume reasonable annual variations. When very large and expensive wind turbines are involved, the design, fabrication and operating costs warrant detailed long-term wind surveys. In the more frequent instance of a relatively small off-the-shelf turbine purchased from a manufacturer, the determination of the available wind energy is likely to be far less precise and practical considerations such as proximity to an existing building, a lake, or a well, generally determine the location of the turbine. In such cases cost-effectiveness is a matter of hope rather than calculation.

Lack of accurate data on long-term wind speed distribution has resulted in numerous instances of disappointing results in terms of actual versus the expected returns on the investment in a wind turbine system. The common practice of using the readily available wind speed data from a nearby airport to estimate power and energy availability is extremely risky since airports are generally located on large flat areas over which wind can run freely while a particular site may be influenced by local hills, buildings and vegetation. Assumptions regarding energy capture which are based on a limited number of measurements at a particular site are risky, but assumptions using extrapolations from nearby site data are little more than guesses. Some of the variables involved in wind speed distributions were outlined in Chapter 3, Energy in the Wind.

Lack of precise data on wind energy data may delay some of the large wind turbine projects in which small miscalculations may result in large financial difficulties, but smaller projects generally must proceed using the best available results. In such cases, it is necessary to consider probabilities within a band of uncertainties so that decisions regarding the feasibility of a project can be made on the basis of a factor of confidence. A convenient model of wind speed distributions, suggested by SERI (1) was used in Chapter 3 to estimate the annual energy capture. In this model, the number of hours, H, in which the wind speed exceeds the value V whose average at that height is V_a, is given by

$$H = a\, d^{-b\lambda^c} \qquad (1)$$

with

$$\lambda = V/V_a$$

$$a = 8766$$

$$b = \pi/4.06$$

$$c = 2.27$$

The above relation does not contain any effects of orientation so that calculations of annual energy capture imply that the wind turbine is omnidirectional. This would be true in the case of horizontal axis turbines with yaw mechanisms which keep them pointed precisely into the wind, or vertical axis turbines which are insensitive to wind directions. A turbine of projected area A, operating at a fixed design power coefficient $C_{P,0}$ would capture an increment of energy dE in the wind speed range λ to $\lambda + d\lambda$. The relation for the incremental energy capture was shown in Chapter 3 to be given by

$$dE = E_0\, bc\lambda^{c+2}\, e^{-b\lambda^c}\, d\lambda \qquad (2)$$

with

$$E_0 = C_{P,0}\rho V_a^3\, Aa/2$$

If a turbine could extract energy at all wind speeds then the annual energy capture would be given by

$$E = E_0 \, bc \int_0^{\infty} \lambda^{c+2} e^{-b\lambda^c} \, d\lambda \qquad (3)$$

The value of the integral is about 1.65 which means that a turbine operating at a constant power coefficient at all wind speeds will capture 65% more energy than would be predicted on the basis of average wind speeds.

Real wind turbines cannot operate at all wind speeds and have variable power coefficients so that the above represents the theoretical upper limit against which performance of wind turbines may be measured. Factors influencing energy capture are:

1. Orientation relative to the wind vector, if the design is not omnidirectional.
2. Wind speeds at which the turbine is started and stopped.
3. Design wind speed if the turbine is operated in a power limited mode.
4. Sensitivity to off-design conditions.
5. Response to transients such as gusts.

From a practical point of view, the most interesting parameter for any wind turbine design is the ratio of the total energy extracted to the total energy available. This, of course, is a function of the variables outlined above, and is as much a measure of performance as the maximum power coefficient which can be achieved. Wind turbines designed to supplement or supplant the power from the electrical grid are forced to operate at fixed rotational speeds in order to generate an alternating current of the correct frequency. Use of converters and inverters is discouraged by the high cost of equipment and generally low overall efficiency. This situation may be changing because of the developments in variable drive synchronous generators (2). Because of the variability of the wind, constant rotational speed wind turbines are forced to operate at off-design conditions almost all of the time. If the angular rate was an entirely free parameter, then without any regard to centrifugal stresses and variations in the blade angle of attack, a propeller turbine could be operated at its peak power coefficient throughout the wind speed range. For an actual turbine the ratio actual energy captured to that which could be captured if the turbine operated at its maximum power coefficient is given by

$$R = E/E_0 = bc \int_0^{\infty} F\lambda^{c+2} e^{-b\lambda^c} \, d\lambda \qquad (4a)$$

$$F = C_P/C_{P,0} \tag{4b}$$

The function F expresses the variation of the power coefficient in off-design conditions.

Wind turbines are analyzed in terms of tip speed ratios which can be achieved by any combination of the radius, angular rate and the wind speed. In energy calculations, the wind speed is the variable and the equations and variables must be recast into a slightly different from. Following the outline given in the discussion of the Savonius rotors, the tip speed ratio for propeller and Darrieus rotors will be defined as

$$X/X_0 = 2Y\lambda^{-1}$$

with

$$Y = \frac{1}{2}\frac{X_d}{X_0}\frac{V_0}{V_a}$$

As before, Y is the design parameter in which X_d may be the design speed while X_0 is the characteristic speed of the turbine.

PROPELLER TURBINES

The annual energy capture of the propeller turbines for which off-design performance was calculated in Chapter 4 is shown in Figure 1, for constant rotational rate operation. There are several significant results which should be noted. Propeller turbines with design tip speed ratio $X_0 = 2$ capture more energy than the generally higher performance turbines with $X_0 = 4$ because of their greater tolerance to off-design operation. Designs biased towards higher wind speeds, $Y = 0.7$ rather than $Y = 0.5$ show significantly greater energy capture ratios. For comparison, the energy capture of a Darrieus rotor using a curve-fit of the power coefficients presented by Ashley (3) is also shown in the Figure. It should be noted that propeller turbines could not be operated over the full range of speeds so that the lower and upper portions of the energy integral should be disregarded, or subtracted from the total.

In general, propeller turbines are designed to start at a certain wind speed and to maintain a constant peak power above a design wind speed until excessive wind speeds demand shutdown. This limits the extraction of power at high wind speeds, but allows the use of a reasonably sized generators. The principal parameter here is the wind speed at which the power output becomes limited.

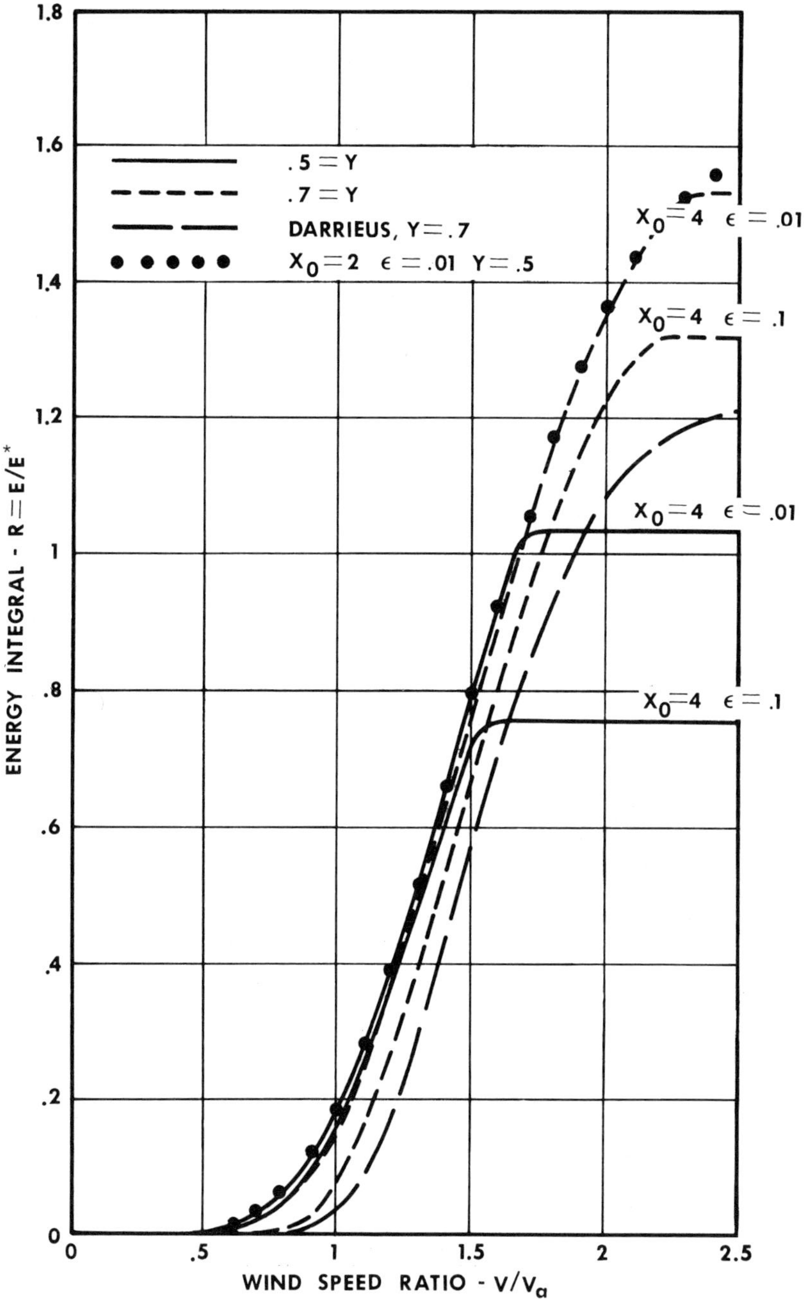

Figure 1. Influence of the design point on the extraction of wind energy

The data in Chapter 4 indicates that some off-design power coefficients of propeller turbines may be approximated by

$$F = C_P/C_{P,0} = 1 - (X/X_0 - 1) = 1 - \left[(\lambda\zeta)^{-1} - 1\right]^2 \qquad (5)$$

with

$$\zeta = V_0/V_a$$

the ratio of the wind speed at which the power output limiting starts, to the average wind speed. In the computer code for the energy integrals constant power is maintained for $\lambda \geqslant \zeta$ by imposing the condition

$$F = \lambda^{-3} \qquad (6)$$

Variations of the total energy integral E_∞ with ζ is shown in Figure 2. Calculated values of the energy integrals for this mode of operation are shown in Figure 3.

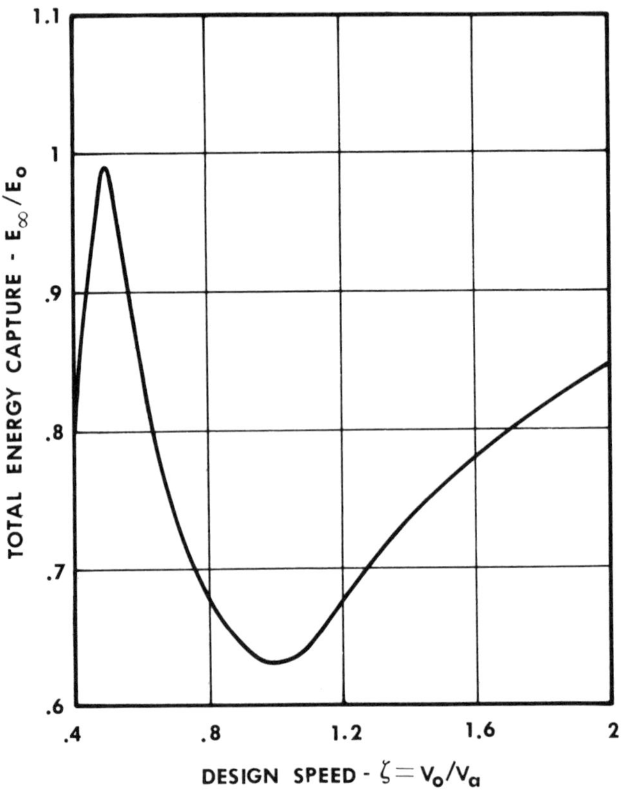

Figure 2. Variation of total energy capture with power limiting speed

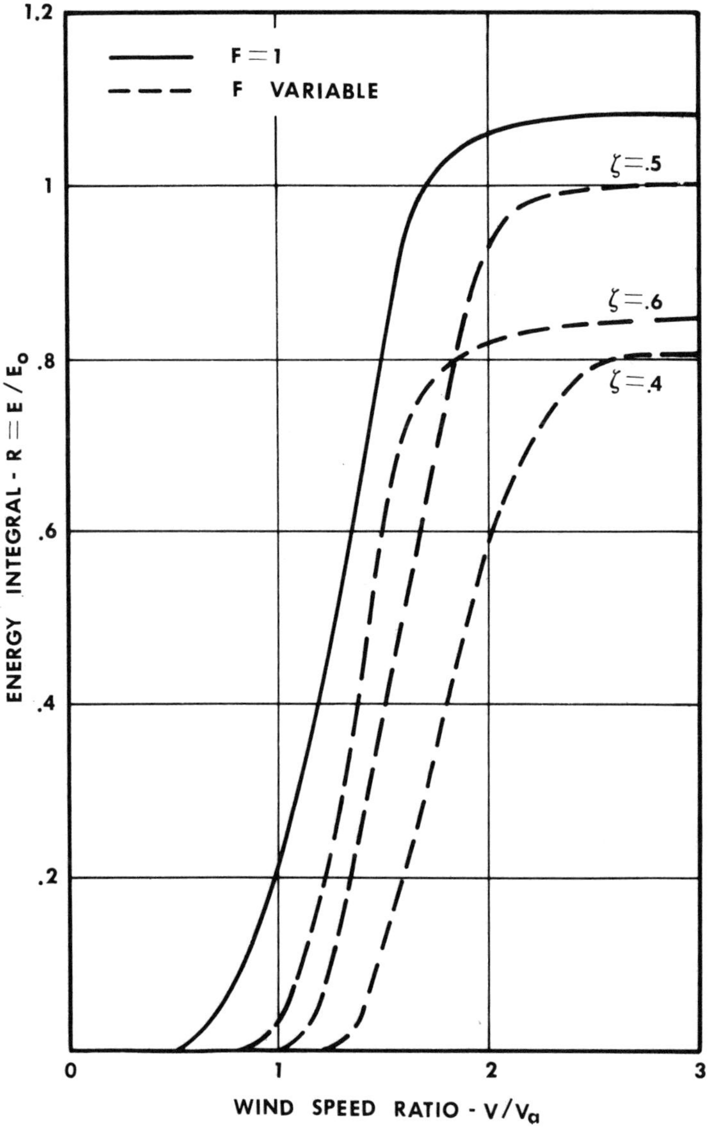

Figure 3. Energy integrals of power limited turbines

The results indicate that highest energy capture can be achieved when the speed at which power limiting is initiated is about 50% of the average wind speed, $\zeta = 0.5$. This is not a practical result because a turbine designed for such low speeds would have to be very large and expensive for its nominal power output. For this reason power limited turbines are designed for ζ of about 1.5 - 2, where the total energy capture is lower, but so is the turbine size and cost.

SAVONIUS ROTORS

Calculations of the energy integrals for Savonius rotors were performed using

$$F = 4Y\lambda^{-1}(1 - Y\lambda^{-1}) \tag{7}$$

The curves shown in Figure 4 indicate that the maximum energy capture is achieved with $0.6 \leqslant Y \leqslant 0.7$. As before, the lower and upper portions of the curves should be disregarded as they represent extreme wind conditions which could not be tolerated by Savonius rotors.

ANNUAL ENERGY ACQUISITION

The results shown above indicate that it is reasonable to assume that an energy ratio $R = E/E_0$ of about unity could be achieved. With an average United States wind speed of 9.5 mph (4.25 m/s) the total annual energy flow is about 38.6 kWhr/ft^2. With a maximum power coefficient of 0.3 the total annual energy capture by a wind turbine would be 11.6 kWhr, or approximately an equivalent of $1 in electric energy value. A 20 ft diameter turbine would therefore capture only about $314 in value of electricity. The transmission and generator combined efficiencies are typically under 80% so that the annual yield of a 20 ft diameter turbine would be only about $250. Such considerations are crucial to the estimates of cost-effectiveness and economic feasibility studies which are addressed in the next chapter.

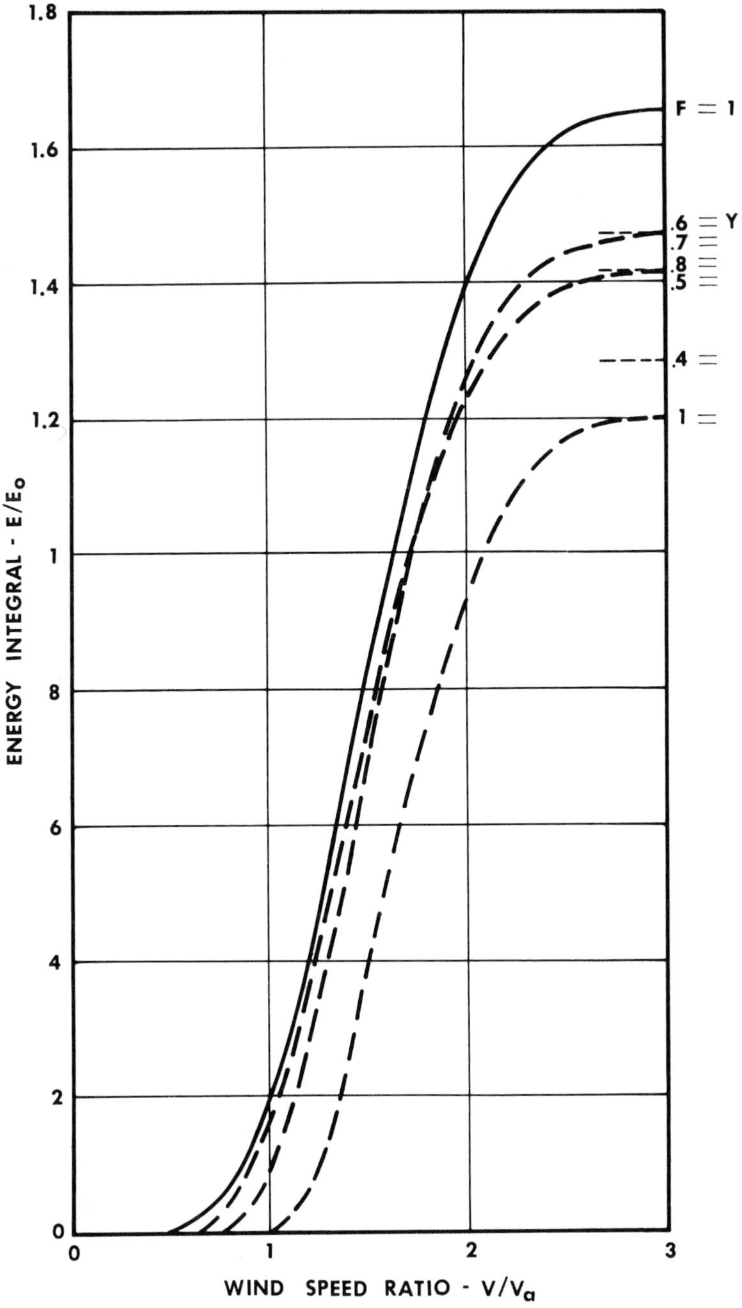

Figure 4. Energy integrals of Savonius rotors

REFERENCES

1. Solar Energy Research Institute, Request for Proposal No. RH-9-8003, Generic Studies of Wind Energy Systems, Oct. 1978.
2. Klass, P. J., "Solid State Power System Use Expands," Aviation Week, Vol. 114, No. 2, Jan 1981, pp. 70-72.
3. Ashley, H., "Some Contributions to Aerodynamic Theory for Vertical Axis Wind Turbines," AIAA Journal of Energy, Vol. 2, No. 2, March-April, 1978, pp. 113-119.

CHAPTER 8

COST OF WIND POWER

Accurate estimates of the true cost of wind power are extremely difficult because very few machines have been produced in significant quantities and scant records exist of the real operating and maintenance costs. Adding to the difficulties are the wide variety of forms of the final output of wind turbine systems and the general lack of detailed installation cost breakdowns.

UTILIZATION OF WIND ENERGY

The mechanical energy produced by the windmill may be used in the following ways:

a. Heating by mechanical or hydrodynamic dissipation
b. Driving of pumps and machinery
c. Generation of electricity

The first method is frequently suggested, but it must be rejected since the high grade energy of the windmill is obtained at a very high cost and it should not be dissipated in heat for the same reason that electrical heaters are the most wasteful means for using electrical energy. The cost of fuel is about 1/3 of the cost of equivalent amounts of electrical energy so the use of windmill power for direct heating could only be contemplated if the cost of producing wind energy could be reduced drastically.

Driving of pumps, or machinery is much more sensible since it involves a direct and efficient transfer of power. The difficulties in driving other types of machinery are due primarily from the variability of windmill power with time of day or season. Only machines without a rigidly fixed schedule could be driven by windmills. Driving of pumps for irrigation or storage for a hydroelectric project is a very attractive possibility which has drawn considerable attention. Power demand of many pumps varies as the cube of their speed so that direct coupling to a windmill operating at its optimum tip speed ratio, but variable rotational speed, represents an excellent match in supply and demand characteristics. Depending on the type of wind turbine, location of the project and the power requirements, considerations of starting torque of the windmill assume various levels of importance.

A large number of mechanical to electrical power conversion schemes have been proposed. Among these are:

a. DC generators with battery storage
b. DC generators with inverters to convert to AC power
c. Variable speed, constant frequency output generators

d. Induction generators
e. Synchronous generators

The first alternative is of limited value because of the 85-90% charging efficiency and the 75-80% discharge efficiency of normal batteries. With realistic accounting for gearbox, control system and charging/discharging efficiencies, it is easy to see that only about 1/2 of the energy produced by the windmill is ultimately available for useful applications. Most of the household and industrial equipment requires AC power so that an expensive inverter with its associated power losses must be added to the system. Economic considerations which include cost of all the equipment, including the batteries, make a battery storage system rather unattractive.

Alternatives b and c allow the windmill to operate at its peak efficiency over a wide range of wind speeds so that a greater fraction of the available wind energy can be captured. These systems are heavier and more expensive than induction or synchronous generators. However, recent developments in electronically operated variable input speed constant frequency output generators (1) appear to have overcome most of the above objections.

Because of the economic necessity of tying larger windmills into the main power grid, it is necessary to generate electricity at a fixed frequency. Induction generators can achieve that with an almost constant speed drive, but synchronous generators require an absolutely constant speed input. This objective can be achieved either by a very stringent control of the windmill rotational speed, or very expensive mechanical or hydraulic constant speed output drives.

It therefore appears that the larger wind turbines must operate at constant rotational speeds using variable pitch blades for speed control while the simpler small machines are more suited to pumping applications, as is done presently with the American farm type windmill. Developments in variable speed constant frequency generators may change the role of the small mills to become sources of supplemental electrical power for individual homes or small establishments.

COMPONENTS AND SYSTEMS

Estimates of cost of power derived from wind turbines are probably the most difficult part of the evaluation process of a wind turbine system. The limited experience with design, fabrication and operation of wind turbines is not adequate for establishing a basis for estimating the costs of future systems. Adding to this problem are the various schemes for augmenting the power of wind turbines which in effect means that for a given power output, large stationary surfaces are exchanged for additional expensive rotating components. Because of the lack of data and agreement of estimating procedures, very little has been done to present concrete data.

It appears from the very limited amount of actual data, that different cost estimating procedures must be used for large wind turbines with ratings greater than say 100 kW and small units with nominal ratings of 1-10 kW. In this section, no attempt will be made to estimate augmented wind turbines since even their strongest proponents (e.g., Gilbert and Foreman (2)) are unable to make any statement beyond an opinion that in some power rating ranges of the augmented turbine, specific power costs will be lower by an unspecified amount than those for conventional turbines. The principal problem in estimating the costs of large wind turbines may be in the complete lack of agreement on the contribution of various major components to the total cost of a wind turbine. An excellent example of that is given by Golding (3) whose compilation of the component costs of four large propeller turbines is summarized here in Table I in the form of ranges to indicate the enormous spread in data.

Table I. Ranges of Costs of Principal Elements
As Percentages of Total System Costs

SYSTEM ELEMENT	RANGE OF PERCENTAGES OF TOTAL SYSTEM COST
Blades	3.0 - 11.2
Hub, Shaft, Nacelle	5.9 - 41.5
Gearbox, Generator	13.5 - 36.8
Control System	4.4 - 10.4
Tower or Supports	7.7 - 31.3
Foundation and Site	9.0 - 31.4
Engineering	3.6 - 14.5

It is interesting to note that Golding's data, which is primarily for European wind turbines and the old U.S. Putnam machine, indicate rotor costs which are one tenth to one fifth of the 43.7% rotor cost contribution given by Eldridge (4) and the 67% and 40% contribution to 100 kW and 1000 kW turbines which are given by Reference 5. Obviously the very large number of variables in wind turbine systems makes it extremely difficult to identify the individual cost elements.

COST ESTIMATING RELATIONS

A very interesting attempt to develop cost algorithms for a broad range of propeller-type wind turbine systems was made by Smith (6) of Smith Wind Systems, Oakland, California who correlated existing data and prediction techniques to arrive at some very simple and very useful relations. In Smith's correlations costs are identified as being dependent on the power rating of the system or on the diameter of the rotor. The power output P in kW is given as

$$P = 0.21D^{2.08} \tag{1}$$

with D being the propeller diameter in meters. The exponent on D is 2.08 rather than 2.0 which would emerge from purely aerodynamic considerations because larger turbines operate at higher Reynolds numbers and thus have higher lift/drag ratios. Also, large generators tend to be somewhat more efficient. The power dependent costs in 1977 U.S. dollars are given in terms of power P in kW by the following formulae:

Nonrecurring costs & fees	$C_{cf} = 95.4P$	(2a)
Foundation costs	$C_f = 64.8P$	(2b)
Nacelle	$C_n = 52.4P$	(2c)
Generator	$C_g = 40.4P$	(2d)
Yaw System	$C_y = 34.0P$	(2e)
Transportation costs	$C_t = 16.0P$	(2f)
Spare parts	$C_{sp} = 14.0P$	(2g)

The costs, in 1977 U.S. dollars, of components which depend primarily on the rotor size are given in terms of the propeller diameter in meters by the relations

Rotor	$C_R = .0184D^{3.43} + .541\ D^{2.8}$	(3a)
Installation	$C_I = 80D^{1.67}$	(3b)
Tower	$C_T = 11D^{2.26}$	(3c)

The two terms in the rotor cost equation represent the material and labor costs respectively. These relations were used by Smith (6) in a study spanning a two orders of magnitude range of power outputs. However, in view of the sparseness of data for such a wide range of turbine sizes, the results obtained through the use of the above relations must be treated with caution. As is frequently found in engineering until better data are found the simple relations must be used, provided that their limitations and possible errors are clearly understood.

COSTS OF COMMERCIAL SYSTEMS

Actual costs of some selected wind turbines currently available commercially are shown in Table II. Only a representative sample is presented here, and in all cases where a range of prices was quoted the higher figure is used. While it is possible to buy windmills for as little as $1800/kW it appears more likely that the cost will be twice that. Direct comparisons among the data in Table 1 are difficult since detailed specifications of the turbine and auxiliary equipment are not available. For a more detailed listing the compilation of Hunt (7) may be consulted.

Table II. Cost of Propeller Wind Turbines

MANUFACTURER	MODEL	BLADE Diameter-ft	RATED Power-kW	COST $	COST, $ kW
AEOLIAN ENERGY	AEO-20	20	4	10,000	2,500
	AEO-32	32	10	17,750	1,775
BERGEY	1000-S	8.3	1	3,000	3,000
DUNLITE	2000	13	2	7,700	3,850
ENERTECH	4000	20	4	16,000	4,000
JACOBS	10-kW	23	10	20,000+	2,000+
SENCENBAUGH	1000-14	12	1	3,600	3,600
WINDWORKS	10	33	9	29,500	3,400
ZEPHYR WIND DYNAMIC	WIND TILTER	7.4	6.5	1,500	3,000

COST OF ENERGY

Calculations of the total cost of energy must include not only the initial costs of equipment and installation but also all costs associated with operations and maintenance. Methods of calculating costs range from trivially simple to very large computer codes which generate large quantities of numbers and dazzling praphical displays. The true value of the extremely elaborate cost computations will remain in question until sufficient data are accumulated to provide a solid basis for predictions.

A very simple method which is useful for rapid estimates considers only the cost of money and operating and maintenance costs. It is customary to present the cost of wind turbines in terms of cost/kW of nominal power. With such a figure, it is possible to make a very crude estimate of the cost of power. It must be realized that the specific power cost is a function of the rated wind speed, but this is not accounted for here. A turbine with a power rating of P will produce an annual energy output of

$$E^* = PR8766 \tag{4}$$

with R being the energy integral which was given in the previous chapter. If the cost of the turbine is C_0 on a per kW basis, the annual cost of money is c_m and the annual maintenance cost is a fraction c_n of the initial cost, then the total cost for the life of N years of the machine will be approximately

$$C_E = \left(C_0 P\ 1 + c_m N + c_n N\right) \tag{5}$$

With all the uncertainties in C_0 sometimes there is no reason to employ more refined accounting procedures. The cost of energy will therefore be

$$C_E = \frac{\left(N^{-1} + c_m + c_n\right)}{8766R} C_0 \tag{6}$$

with

$$C_0 = 3600/\text{kW}$$
$$c_m = 0.18$$
$$c_n = 0.06$$
$$N = 20 \text{ years}$$
$$R = 1.0$$

The cost of electricity is found to be \$0.119/kWhr. If such figures could be believed then wind power would indeed be quite attractive. This number is approximately equal to that estimated by Coty and Vaughn (8) on the basis of a production run of 200 units of 180 ft diameter, 200 kW rated turbines. A more likely cost figure is about 30-40¢/kWhr when realistic fabrication and maintenance costs are included.

A much more elaborate method for computing the cost of energy delivered by propeller turbines was presented in Reference 9. The basic equation is

$$C_E = \left(C_T f_T + C_L f_L + C_R f_R + C_0 f_0\right)/E \tag{7}$$

with C_E amd E being the costs and energy output respectively calculated on an annual basis. The various components of costs are:

C_T = Cost of turbine equipment

C_L = Cost of land

C_R = Cost of periodic replacements

C_D = Cost of operation and maintenance

The multiplying factors f, are as follows

f_T = Equipment fixed rate charge = 0.18

f_L = Land fixed rate charge = 0.15

f_R = Periodic levelizing factor

= $C_{RF}(1 + e)/1 + r)^i$

with

C_{RF} = Capital recovery factor = 0.089

e = Price escalation rate = 0.06

e = Discount rate = 0.08

i = Years after installation

f_0 = Annual levelizing factor = 2.0

The above relations represent the recommended NASA accounting procedures for the Mod-6H wind turbine and are probably adequate for all preliminary cost estimates.

An approximate idea of the cost of electricity generated by wind power may be obtained by considering a 10kW wind turbine costing \$36,000 with installation, power limited at wind speeds greater than 30 mph. If the machine is located on an acre of land worth \$1000 in which the annual average wind speed is 22.5 mph (Note: U.S. average is about 9.5 mph) then the cost of energy can be readily estimated. Assuming that R = 0.8 the annual energy output is

$$E = 10.8766 \ \frac{22.5}{30}^3 \ 0.8 = 29,585 \text{ kWhr} \qquad (8)$$

Annual cost of equipment = 0.18 x 36,000 = \$6,480. If it is assumed that the annual replacement cost is 5% of the original cost then in the third year cost of replacements = 1800 f_R = \$135. If the annual cost of operation and maintenance is taken as 6% of the original cost then cost of O & M = 2 x 0.06 x 36,000 = \$4320, Cost of land = .15 x 1000 = \$150. The total annual cost is therefore found to be \$11,085. The cost of electricity is therefore \$0.375/kWhr, or about 4 times the current U.S. rate. It must be noted that if the turbine was located at a site with an average wind speed of 15 mph then the annual energy output would be only 10,958 kWhr and the cost of electricity would be \$1.01/kWhr.

It is realized that cost estimates of the type presented above are subject to some uncertainties and thus are more of an indication rather than a true estimate. Nevertheless, these are the best estimates available at present and if they are used judiciously in the evaluation of the economic feasibility of wind turbine systems then considerable gains will be made in the development of a rational approach to the evaluation of the cost effectiveness of exploitation of wind energy.

REFERENCES

1. Klass, P. J., "Solid State Power System Use Expands," Aviation Week, Vol. 114, No. 2, Jan. 1981, pp. 70-72.
2. Gilbert, B. L., and Foreman, K. M., "Experimental Demonstration of the Diffuser Augmented Wind Turbine Concept," AIAA Journal of Energy, Vol. 3, No. 4, July-August 1979, pp. 235-240.
3. Golding, E. W., The Generation of Electricity by Wind Power, Halstead Press, New York, 1977, p. 265.
4. Eldridge, F. R., "Wind Machines," Report NSF-RA-N-75-051, National Science Foundation, 1975.
5. Considine, D. M., Ed., Energy Technology Handbook, McGraw-Hill, New York, 1977, p. 6-173.
6. Smith, D., Private communication.
7. Hunt, V. D., Windpower, Van Nostrand Reinbold Co., New York, 1981.
8. Coty, U., and Vaughn, L., "Implementation Issues of Wind Energy," Proceedings of AIAA/EEI/IEEE Conference on New Options in Energy Technology, 1977.
9. NASA Request for Proposal, RFP-182390-Q, Advanced, Multi-Purpose Medium-Scale Wind Turbine, 20 Feb. 1980.

CHAPTER 9

PRELIMINARY DESIGN PROCEDURES

Preliminary design and analysis of wind turbine systems is complicated by the fact that required functions may be performed by several distinctly different wind turbine designs coupled to different coutput systems. In order to realize the optimum wind systems, it is necessary to define precisely the system characteristics and to know in considerable detail the principal operating and cost features of the most common designs such as the propeller, Darrieus and Savonius rotors. If the turbine itself is to be built rather than purchased from a manufacturer then it is necessary tohave a firm grasp of applied aerodynamics, wind turbine theory, turbine testing procedures, and scaling relations if the model and prototype are of different sizes. Details of the preliminary design procedure are strongly dependent on the size of the system, but the basic line of logical reasoning and engineering analysis applies to all projects. The principal tasks are as follows:

Preliminary Feasibility Study

Requirements	-	Identify requirements
Alternatives	-	Analyze competitive sources of energy such as electric, steam, engine, or watermill systems
Site selection	-	Preliminary estimates of the potential of possible sites
Turbine size	-	Estimate the gross size of the turbine on the basis of reasonable power coefficients
Feasibility	-	Determine feasibility on the basis of costs and the value of energy

Overall System Considerations

Demand	-	Power, minimum, maximum, and daily and seasonal variations, annual energy requirement
Power Output	-	AC or DC generators, pump drive, operation ofmachinery, energy storage
Load Characteristics	-	Constant or variable speed, power-speed relations, starting torque
Site	-	Specified or open, wind speed distribution, available wind energy
Economic Feasibility	-	Preliminary cost estimate, cost of energy

Preliminary Engineering Analysis

Turbine Type Selection	- Propeller, Multiblade, Darrieus, Savonius, Combination of above
Sizing	- Peak power, average power, annual energy output
System Configuration	- Tower height and type, gearbox and transmission, connection to load
Control Requirements	- Start and stop, regulation of speed, regulation of output

Performance Estimates

Turbine Characteristics	- Peak power coefficient, off design performance, transmission efficiency
Power Output	- Detailed calculations of daily, monthly and seasonal power output
Energy Output	- Daily, seasonal and annual energy output

Mechanical Design

Blades	- Aerodynamic surfaces, stress levels, gust overloads, fatigue considerations, materials selection
Gearbox	- Type of geared drive, couplings
Transmission	- Generator type, mechanical output, fluid drive
Controls	- Dynamics of the system, control system characteristics, control system selection
Tower and Foundation	- Tower type and height, foundation design, stresses, fatigue life
Packaging	- Supports, accessibility, weather protection
Environmental Considerations	- Noise, TV interference, appearance, hazard to aircraft, attraction to vandalism, hazard to surroundings in case of mechanical failure

Costs

Initial Costs	- Analysis and design, fabrication, assembly and test, transportation, final checkout
Operations	- Routine checks, analysis
Maintenance	- Routine service, routine replacements, repairs
Life	- Major components, subsystems

Operation and Maintenance

Components	-	Accessibility, service intervals
Controls	-	Checks, adjustments
Replacements	-	Routine, failures

The above outline is a rough indication of the magnitude of the preliminary design process involved in the development of a wind turbine driven system. Obviously, if a wind turbine is purchased then that aspect of the problem is eliminated. On the other hand many of the items shown above can only be determined after several iterations involving various components of the system. Not shown in the above are environmental conditions which the wind turbine system must be able to withstand. To illustrate representative requirements the following are reproduced from Reference 1, and thus represent current standards.

Other Factors

Temperature	-	-40^oC to $+49^oC(-40^oF$ to $+120^oF)$ ambient air
Rain	-	10cm/hr (4 inches/hour)
Hail	-	2.5cm (1 inch) diameter, 20m/sec (45 mph) velocity
Ice (Glaze)	-	5cm (2 inches) thickness 961 kg/cum (60 lb/cuft) on all external surfaces
Snow	-	Blades 100kg/sqm (21 lbs/sq ft)
Humidity, Sand/ dust, Salt, Spray, Fungus		exposure equivalent to MIL-STD 210B for exposed or sheltered ground equipment as applicable
Fauna	-	Insects and 1814gm (4 lb) birds) moving at 16m/sec (35 mph)
Solar radiation	-	98,500 cal/m^2hr (363 btu/ft^2/hr) 4 hours daily
Altitude	-	Sea level to 7000 ft

Vandalism

The system shall be designed to protect against vandalism. Normal design practices shall be utilized; however, consideration shall be given to damage caused by rifle fire and preventing entry of unauthorized personnel into secured areas (i.e., areas containing wind turbine controls, rotating machinery, high voltage, etc.)

Planning and design of a wind turbine system requires expertise in so many fields that only teams of engineers can undertake larger projects. Smaller projects, particularly those based on purchased wind turbines may be analyzed using the data presented in this book. Supplementary material on evaluation of wind turbine generating systems may be found in Reference 2. Investment in time and effort to understand some of the basic features of wind turbine systems may ultimately prove to be the optimum approach when hard decisions regarding wind energy projects are necessary.

REFERENCES

1. NASA, Request for Proposal - Advanced Multi-Purpose, Medium-Scale Wind Turbine, Feb. 1980.
2. Wortman, A., "Guidelines for Evaluating Wind Turbine Generating Systems," Report AWD-R-81-0201, Feb. 1981.

LIST OF SYMBOLS

a	–	axial interference factor radius of a hill total annual hours
a'	–	tangential interference factors
$\ddot{a}$	–	wing effectiveness factor, ~ 0.9
A	–	amplitude of gusts streamline function, ψ/u_a turbine projected area
A_R	–	aspect ratio
b	–	constant in the wind speed distribution equation
c	–	section chord exponent in the wind distribution equation
c_p	–	specific heat at constant pressure
c_{pr}	–	diffuser exit pressure coefficient
c_P	–	local power coefficient
c_T	–	torque coefficient
c_v	–	specific heat at constant volume
C_D	–	drag coefficient
C_{Fx}	–	axial force coefficient
C_L	–	lift coefficient
$C_{L\alpha}$	–	lift slope coefficient
C_{Mx}	–	moment coefficient
C_P	–	power coefficient
C_T	–	torque coefficient
D	–	drag
e	–	efficiency
E	–	energy
E_0	–	annual energy at V_a
E^*	–	annual output of a rated turbine
f	–	geometrical parameter, $C_{D,0}\ell/2R$
f_a	–	axial force
F	–	off-design performance function
h	–	height above ground
h_o	–	hub height

H	-	enthalpy change hours
I	-	impulse function
ℓ	-	mixing length paddle turbine arm length
L	-	lift
L_0	-	magnitude of lift
m	-	exponent on speed-height relation mass
$\dot{m}$	-	mass flow rate
N	-	number of blades number of years of service
p	-	pressure
p'	-	induced pressure at the disk
P	-	power
q	-	dynamic pressure flow speed
Q	-	average velocity defect in the wake
r	-	local radius
R	-	energy capture turbine radius
R_A	-	ideal capture ratio
R_E	-	energy fraction yawing to fixed
R_p	-	energy ratio, gusts to steady
T	-	period of gusts tangent vector temperature torque
u*	-	friction velocity
U	-	flow velocity
v	-	local velocity
V	-	wind speed
w	-	radial velocity magnitude of relative velocity
W	-	complex potential, $\phi + i\psi$ relative velocity
x	-	horizontal coordinate local tangential speed ratio

X	-	tangential speed ratio
X_0	-	turbine tip speed ratio
Y	-	turbine design operating parameter
z	-	dimensionless radius, r/R vertical coordinate
z_0	-	equivalent surface roughness height

Greek

α	-	angle of attack
γ	-	isentropic exponent
Δ	-	paddle turbine geometrical parameter
ζ	-	pitch angle power limiting wind speed ratio
ϵ	-	drag/lift ratio
λ	-	blade loading coefficient wind speed ratio, V/V_a
θ	-	polar angle
ν_t	-	eddy diffusivity coefficient
ρ	-	air density
σ	-	solidity
τ_0	-	shear stress at earth's surface
τ_{xz}	-	shear stress
ϕ	-	flow angle potential function
ψ	-	stream function
ω	-	turbine angular rate
Ω	-	turbine design angular rate

Subscripts

a	-	average
b	-	base
B	-	average base
d	-	design
i	-	inlet
P	-	power

r	-	radial reference relative
t	-	total
0	-	free stream optimum zero lift for drag
1,2...	-	station
θ	-	tangential

The large number of symbols necessitated some duplication without, however, any ambiguities because similar symbols appear in different parts of the book.